EVALUATION OF GREENHOUSE GAS EMISSIONS FROM SEPTIC SYSTEMS

by:

Harold L. Leverenz, P.E., Ph.D.
George Tchobanoglous, P.E., Ph.D.
Jeannie L. Darby, P.E., Ph.D.
University of California, Davis

2010

WERF

The Water Environment Research Foundation, a not-for-profit organization, funds and manages water quality research for its subscribers through a diverse public-private partnership between municipal utilities, corporations, academia, industry, and the federal government. WERF subscribers include municipal and regional water and wastewater utilities, industrial corporations, environmental engineering firms, and others that share a commitment to cost-effective water quality solutions. WERF is dedicated to advancing science and technology addressing water quality issues as they impact water resources, the atmosphere, the lands, and quality of life.

For more information, contact:
Water Environment Research Foundation
635 Slaters Lane, Suite G-110
Alexandria, VA 22314-1177
Tel: (571) 384-2100
Fax: (703) 299-0742
www.werf.org
werf@werf.org

This report was co-published by the following organization.

IWA Publishing
Alliance House, 12 Caxton Street
London SW1H 0QS, United Kingdom
Tel: +44 (0) 20 7654 5500
Fax: +44 (0) 20 7654 5555
www.iwapublishing.com
publications@iwap.co.uk

Library of Congress Catalog Card Number: 2010903239

IWAP ISBN: 978-1-84339-616-1/1-84339-616-5

University of California, Davis

The research on which this report is based was developed, in part, by the United States Environmental Protection Agency (EPA) through Cooperative Agreement No. X-830851-01 with the Water Environment Research Foundation (WERF). However, the views expressed in this document are not necessarily those of the EPA and EPA does not endorse any products or commercial services mentioned in this publication. This report is a publication of WERF, not EPA. Funds awarded under the Cooperative Agreement cited above were not used for editorial services, reproduction, printing, or distribution.

ACKNOWLEDGMENTS

The assistance and feedback from the WERF Program Directors and Project Subcommittee on this project are acknowledged gratefully. Special appreciation is extended to Rebecca Siren, Darrell Creeks, and Jeff Pulfer at the Georgetown Divide Public Utility District and the Auburn Lake Trails community. Assistance and support from Orenco Systems, Rodney Ruskin at Geoflow, Pete Munoz at Natural Systems International, and Barton Kirk at SEEDS was invaluable. At UC Davis, the assistance of Yacov Assa, Heather Dang, Cynthia Kallenbach, and Ugo Eke in the laboratory and field testing, their contribution to this project is greatly appreciated.

Research Team

Co-Principal Investigators:

Harold L. Leverenz, P.E, Ph.D.
George Tchobanoglous, P.E, Ph.D.
Jeannie L. Darby, P.E, Ph.D.
University of California, Davis

Project Team:

Libia R. Diaz-Valbuena, M.S.
Christopher D. Cappa, Ph.D.
William R. Horwath, Ph.D.
University of California, Davis

WERF Project Subcommittee

Water Environment Research Foundation Staff

ABSTRACT AND BENEFITS

Abstract:

This study determined the emission rates of greenhouse gases (GHGs) from individual onsite septic systems used for the management of domestic wastewater. A static flux chamber method was used to measure the emission rates of methane, carbon dioxide, and nitrous oxide gases from eight septic tanks and two soil dispersal systems. A technique developed for the measurement of gas flow and concentration at clean-out ports was used to determine the mass flow of gases moving through the household drainage and vent system. There was general agreement in the methane emission rates for the flux chamber and vent system methods. Several sources of variability in the emission rates were also identified.

The septic tank was the primary source of methane, whereas the soil dispersal system was the principal source of carbon dioxide and nitrous oxide emissions. Methane concentrations from the soil dispersal system were found to be near ambient concentrations, similarly negligible amounts of nitrous oxide were found in the septic tank. All emissions originating in the soil dispersal system were discharged through the building vent as a result of natural, wind-induced flow. The gaseous emission rate data were determined to be geometrically distributed. The geometric mean and standard deviation (s_g) of the total atmospheric emission rates for methane, carbon dioxide, and nitrous oxide based on samples from the vent system were estimated to be 10.7 ($s_g = 1.65$), 335 ($s_g = 2.13$), and 0.20 ($s_g = 3.62$) g/capita·d, respectively. The corresponding total anthropogenic CO_2 equivalence (CO_2e) of the GHG emissions to the atmosphere, is about 0.1 tonne CO_2e/capita·yr.

Benefits:

- Provides methods to determine the GHG emission rates from septic tanks, venting systems, and soil dispersal systems.
- Improves upon the estimation of GHG emission rates from septic tank systems.
- Provides the atmospheric emission rate values for future GHG inventories from septic tank systems in California.
- Examines the GHG generation pathways in typical septic tank system.
- Identifies sources of variability in the GHG emission rates that can be used as a basis for future studies.

Keywords: Onsite wastewater treatment system, anaerobic, flux chambers, greenhouse gases, methane, carbon dioxide, nitrous oxide, septic tank, vent system, soil dispersal system.

TABLE OF CONTENTS

LIST OF TABLES

LIST OF FIGURES

LIST OF ACRONYMS

AB	Assembly Bill
ALT	Auburn Lake Trails
BOD_u	Ultimate carbonaceous oxygen demand
COD	Chemical oxygen demand
ECD	Electron capture detector
FC	Flux chamber
FTIR	Fourier transform infrared
GC	Gas chromatograph
GDPUD	Georgetown Divide Public Utility District
GHG	Greenhouse gas
GWP	Global warming potential
HRT	Hydraulic residence time
IPCC	Intergovernmental Panel on Climate Change
LCA	Life cycle assessment
MB	Mass balance
M_g	Geometric mean
mg/L	Milligrams per liter
OD	Oxygen demand
OLR	Organic loading rate
ppb	Parts per billion
ppm	Parts per million
SRB	Sulfate reducing bacteria
s_g	Geometric standard deviation
SRT	Solids retention time
TDLSS	Tunable diode laser spectroscopy
UASB	Upflow anaerobic sludge blanket
UPC	Uniform plumbing code
USDA	United States Department of Agriculture
U.S. EPA	United States Environmental Protection Agency
VFA	Volatile fatty acid
VSS	Volatile suspended solid
WWTP	Wastewater treatment plant

Executive Summary

Methane has been identified as a potent greenhouse gas (GHG), with an equivalent effect 25 times that of carbon dioxide (IPCC, 2007). Based on the IPCC methodology, the U.S. EPA (2009) has determined that a significant amount of the methane emissions associated with wastewater originate from onsite septic tank systems due to the large number of individual septic systems now in use and the high methane emission rates predicted using the IPCC method. However, the actual data currently available on the emission of methane from septic tank systems are insufficient to produce an accurate greenhouse gas inventory for these systems. Thus, the principal objective of this research was to obtain more accurate data on GHG emissions from conventional septic tank systems, with a focus on methane emissions. To accomplish the objective, this project consisted of a literature review, construction of flux chambers, development of sampling techniques and protocols for gas sampling from septic tank system and ventilation systems, identification and selection of field sites, collection of gas samples, data analysis, and estimation of the GHG emission rate values.

Methodology

For this study, flux chambers and a corresponding sampling methodology were used to measure the GHG emission rates from conventional septic tank systems. The flux chamber method was also used for gas sampling of soil dispersal systems. In addition, a sampling device and methodology were developed to measure the mass flow of GHGs through drainage and ventilation systems. Based on the flux chamber and vent sampling data, emission rates of methane, carbon dioxide, and nitrous oxide from septic tank systems were estimated.

Results

In total, eight septic tanks were sampled for the production of gases from the tank contents using flux chambers. While methane production is attributed to anaerobic reactions occurring primarily in the sludge layer, carbon dioxide emissions result from anaerobic, facultative, and aerobic reactions. Methane and carbon dioxide were the primary GHGs found in emissions from the septic tank, while carbon dioxide and nitrous oxide were the primary GHGs from the soil dispersal system.

The septic tank methane flux values determined by the flux chamber method were found to be log-normally distributed, with a geometric mean (M_g) of 11.0 g/capita·d and geometric standard deviation (s_g) of 2.50. The geometric mean of the methane flux values from individual septic tanks was found to range from 6.3 (s_g = 1.40) to 17.9 (s_g = 1.94) g/capita·d, excluding results from one site that had the septic tank pumped recently. A summary of the methane emission rate values found in the literature and measured values using the flux chamber from this study are presented in Table ES-1.

The average rate of methane emission measured with the flux chamber is not in agreement with the IPCC (2007), Winneberger (1984), and Sasse (1998) models. One reason for the difference between the measurements presented here and those determined from the IPCC (2007) model is that the IPCC (2007) model assumes that half of the influent COD to the septic tank is converted anaerobically. Further, the fate of organic matter present in septage and septic tank effluent is not accounted for clearly. It is likely that the measurements made by Winneberger (1984) are high because not enough samples were obtained to establish a statistical distribution. The values used in the Sasse (1998) model were based on statistics from septic tanks located in developing countries, which operate at higher temperatures and loading rates, resulting in higher methane emissions.

Table ES-1. Summary of Methane Emission Rates Including the Flux Chamber Method.

Method	Year	Methane estimate ($g\ CH_4$/capita·d)
Kinnicutt et al.	1910	10.1[a]
Winneberger	1984	14 to 18[a]
COD loading	2009	11[b]
IPCC	2007	25.5[c]
Sasse	1998	18[d]
This study	2009	11.0 (s_g = 2.50)[a,e]

[a] Measured value.
[b] Calculated value assuming that 40 % of solids are removed as septage.
[c] Calculated value assuming that half of the influent COD is converted anaerobically.
[d] Calculated value assuming 25 % CH_4 dissolved.
[e] Geometric mean and standard deviation as determined using flux chamber method, this study.

The estimated septic tank emission rates reported in terms of CO_2-equivalent (CO_2e) emissions are summarized in Table ES-2.

Table ES-2. Comparison of GHG Emission Rates as CO_2e from the Septic Tank and Vent Average Measurements.

	Geometric mean emission rate value (g/capita·d)			Carbon dioxide equivalent emissions (tonne CO_2e/capita·year)	
Compound	Septic tank	Septic system[b]	GWP[a]	Septic tank	Septic system[b]
Methane	11.0	10.7	21	0.084	0.082
Nitrous oxide	0.005	0.20	310	0.00057	0.023
Carbon dioxide	33.3	335	1	0.012	0.12
Total GHG emissions				0.096	0.23
Total anthropogenic emissions[c]				0.085	0.10

[a] GWP for a 100 year horizon IPCC (1996).
[b] As determined from vent system sampling.
[c] Biogenic carbon dioxide is not included in GHG inventories (U.S. EPA, 2009).

Major Findings

The principal findings from this research are:

- The geometric mean of the total emission rates for methane, carbon dioxide, and nitrous oxide based on samples from the vent system were 10.7 (s_g = 1.65), 335 (s_g = 2.13), and 0.20 (s_g = 3.62) g/capita·d, respectively.
- The CO_2e of the methane emission rates to the atmosphere, as determined from the vent system data in this study, is about 0.10 tonne/capita·yr. Based on the current estimated per capita CO_2e emission rate for the United States (i.e., 23 tonne CO_2e/capita·yr; U.S. EPA, 2010), the septic tank accounts for about 0.5 percent of the total per capita emission.
- The septic tank methane flux values determined by the flux chamber method were found to be log-normally distributed, with a geometric mean (M_g) of 11.0 g/capita·d and geometric standard deviation (s_g) of 2.50. Similarly, the values of M_g for carbon dioxide and nitrous oxide were 33.3 (s_g = 2.73) and 0.005 (s_g = 4.35) g/capita·d, respectively.

Other Findings

Other findings from this research are:

- There was general agreement for methane gas emission rates determined with theoretical models and measured with the flux chamber and vent methods.
- There is considerable variability in the methane gas fluxes from tank to tank and sample to sample.
- The hardness of the water supply appears to influence the overall flux of carbon dioxide, with soft water systems having higher carbon dioxide gas fluxes.
- A correlation between the GHG emission rates from the septic tank and the liquid temperature was not observed, perhaps as a result of the limited duration of the study.
- The presence or absence of a scum layer had no discernable influence on emission rates of methane and carbon dioxide.
- The carbon (measured as COD) in the septic tank effluent discharged to the soil dispersal system was approximately equal to the amount of carbon being vented (measured as carbon dioxide) from the soil dispersal system.
- Methane generated during the anaerobic digestion process was found in both gaseous and aqueous forms but no relation was found between the gas and liquid phase concentrations.
- Nitrous oxide emission rates from septic tanks were found to be negligible; however nitrous oxide was detected in the gases vented from the soil dispersal system.
- Essentially no gas emissions from the soil surface were measured using flux chambers placed above the soil dispersal system trenches.
- Using the mean methane emission value measured for the septic tanks from this project, an MCF value of 0.22 was calculated.

Recommendations

Based on the findings of this research, the following recommendations are proposed for future studies of greenhouse gas emissions from septic tank systems:

- Due to the uncertainty in several operational parameters, such as temperature and water hardness, and their influence on the production of septic tank gases, further study in other regions of the country is recommended.
- In this study, only direct GHG emissions from operational septic systems were evaluated. A follow up study quantifying the GHG emissions associated with various septage management practices is needed.
- The soil dispersal systems in this study were well drained and did not have any ponding. The evaluation of soil dispersal systems in other soil types and at different stages of ponding is needed. In addition, studies on alternative soil dispersal systems, such as pressure and drip irrigation systems, are needed.
- Only conventional septic tank systems were evaluated in this study. Additional research is needed to quantify gas emissions from alternative onsite wastewater systems, such as natural treatment systems, packed bed filters, and other aerobic treatment processes.
- A study is needed to evaluate the development of methanogenesis in septic tanks from the time of start-up, with and without inoculation.
- The methods for sampling of gases from ventilation systems should be further developed and refined.
- Further study is needed to develop technologies for the control of GHG emissions from wastewater systems is needed, including soil-based filtration and stand-alone biofiltration processes. Integration of these types of control systems may require slight modifications to the aspects of the building code related to septic tank ventilation systems.
- A study is needed to determine the correlation between the GHG emissions and the septic tank influent quality and loading.
- Further work should be conducted to quantify GHG emissions from all types of wastewater management systems so that accurate models can be developed.

CHAPTER 1.0

INTRODUCTION

This study was conducted to determine the emission of greenhouse gases (GHGs) from conventional septic tank systems used for the management of domestic wastewater. The project background, objectives, approach, and report organization are discussed below.

1.1 Background

Concerns with climate change have led to an effort to reduce the emission of GHGs, especially in the state of California, which has enacted regulations related to GHG inventory and mitigation (i.e., AB 32). Methane has been identified as a potent greenhouse gas; equivalent to 25 times that of carbon dioxide. Similarly, nitrous oxide has a reported potency factor of 298 times that of carbon dioxide (IPCC, 2007). Using the IPCC methodology, the U.S. EPA (2009) has determined that a majority of the methane emissions associated with wastewater originate from conventional septic tank systems, due to the large number of individual septic systems now in use. However, the actual data currently available on the emission of methane from septic systems are insufficient to produce an accurate GHG inventory for these systems.

In the IPCC (2007) method used currently to compute GHG emissions from septic tank system it is assumed that half of the influent organic matter is converted to methane. Using the IPCC model, the carbon dioxide equivalent (CO_2e) of the methane released from individual domestic septic tanks is about 0.24 tonne CO_2e/capita·yr. However, based on historical studies of methane emissions from septic tank systems, the value is estimated to range from about 0.09 to 0.16 tonne CO_2e/capita·yr. In addition, these estimates only consider emissions from the septic tank and do not account for emissions from the soil dispersal systems. Thus, there is a need to develop a more detailed model of the production of GHGs from septic tank systems.

1.2 Objectives

The principal objective of the research was to obtain more accurate data on GHG emissions from septic tank systems, focusing on methane emissions from conventional septic systems. To achieve the proposed objective the following activities were performed:

- Extensive literature review on septic tanks and gas formation pathways.
- Construction of flux chambers and development of a method for gas sampling from septic tanks.
- Development of sampling techniques and a protocol for gas sampling from septic system drainage and ventilation piping.
- Identification and selection of appropriate field sites for collection of study data.

- Collection of gas samples from the septic tank liquid surface, vent system, and soil dispersal system over a six-month period.
- Analysis of results and determination of estimated GHG emission rate values.

1.3 Project Approach

Methods for the measurement of GHG emissions from soil-plant ecosystems using flux chambers are well established; however, little information is available on gas flux measurements from the liquid/solid surface of septic tanks. Hence, for this research a flux chamber design, based on the design used for soil-based measurements, was constructed and tested, along with the development of a corresponding sampling methodology. The flux chamber method was also used for gas sampling of the soil dispersal system. In addition, a sampling device and methodology were developed to measure the mass flow of GHGs through the vent system. Methane, carbon dioxide, and nitrous oxide emission rates were measured in this study.

1.4 Report Organization

This report is organized into seven chapters. The first chapter provides an introduction and purpose for this project. A detailed literature review on septic tank systems, characteristics, processes, and GHG emissions is presented in Chapter 2.0. The methods employed during and in support of field monitoring are described in Chapter 3.0. A description of the field sites is presented in Chapter 4.0. The results of the gas flux measurements in the septic tank, venting system and soil dispersal system, a mass balance analysis, and the sources of variability in the gas emission rates are discussed in Chapter 5.0. The implications of the results from the research study are presented in Chapter 6.0. Findings and recommendations for further study are summarized in Chapter 7.0. Calculations and support materials are presented in the appendixes.

CHAPTER 2.0

LITERATURE REVIEW

To assess the potential for the release of greenhouse gases from septic tanks, the characteristics of onsite systems are reviewed in this chapter. The subjects considered in this review include an overview of the development of septic tank systems, the physical characteristics and operation of septic tanks, the fundamentals of the anaerobic processes occurring in septic tanks, and the information that exists on the emissions from septic tanks and other wastewater sources.

2.1 Overview of Septic Tank Systems

The septic tank is one of the oldest units available for the primary treatment of wastewater from decentralized sources. The historical background of the septic tank and its importance in decentralized wastewater treatment systems are discussed in this section. An introductory overview of gaseous emissions from septic tanks is also presented along with a brief description of venting systems and the soil absorption field.

2.1.1 Historical Background

Wastewater from individual buildings and small communities is often managed using onsite wastewater systems when a centralized wastewater collection system is not available. Nearly all onsite wastewater systems incorporate a septic tank for primary treatment of influent wastewater (Crites and Tchobanoglous, 1998). A septic tank is a buried, watertight tank designed and constructed to receive and partially treat raw wastewater (U.S. PHS, 1957; U.S. EPA, 2009). It is estimated that about 25 million septic tanks are currently in use in the United States (U.S. EPA, 2002).

Septic tanks were first reported as wastewater treatment systems in the 1860s in France. The Fosse Mouras automatic scavenger was patented in 1881, based on the work of Abbe Moigno and Louis M. Mouras (Dunbar, 1908; Winneberger, 1984). An illustration of the Fosse Mouras septic treatment process is presented on Figure 2-1. The process configuration since that time remains almost unchanged when compared to modern septic systems (Crites and Tchobanoglous, 1998).

2.1.2 Onsite Wastewater Treatment Systems

The key functions of the septic tank are to separate and retain settleable solids (sludge) and floatables (scum) from the incoming wastewater. Subsequently, the treated wastewater is discharged typically into a soil dispersal system, also known as a leach field. The captured solids are retained in the septic tank and undergo a passive (naturally occurring and uncontrolled) anaerobic digestion (Tchobanoglous and Schroeder, 1985). The combination of septic tank and leach field, shown on Figure 2-2, is the most commonly used onsite wastewater treatment system. Other types of primary treatment processes used in decentralized wastewater systems include Imhoff tanks, anaerobic baffled reactors, and upflow anaerobic sludge blanket reactors (Crites and Tchobanoglous, 1998; U.S. EPA, 1999). However, these alternative primary treatment processes are not used commonly due to their more complex construction and operation. As shown on Figure 2-2, the septic tank is connected to a building through subsurface drainage pipes and the treated wastewater (septic tank effluent) is discharged to the soil dispersal system (U.S. EPA, 2000).

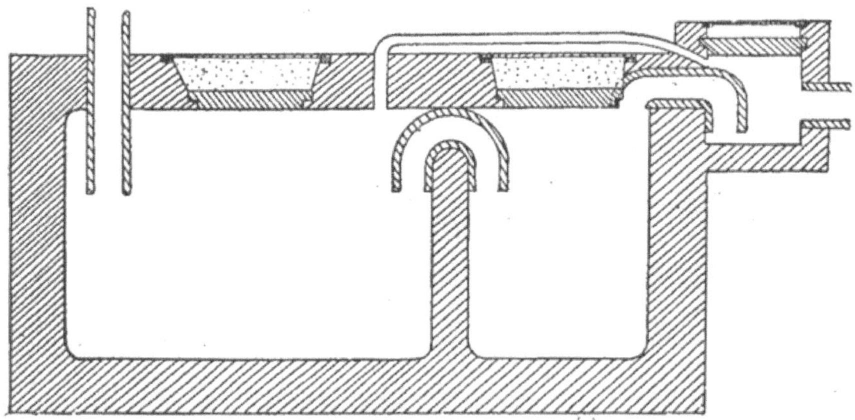

Figure 2-1. Illustration of the Mouras Automatic Scavenger. Adapted from Dunbar, 1908.

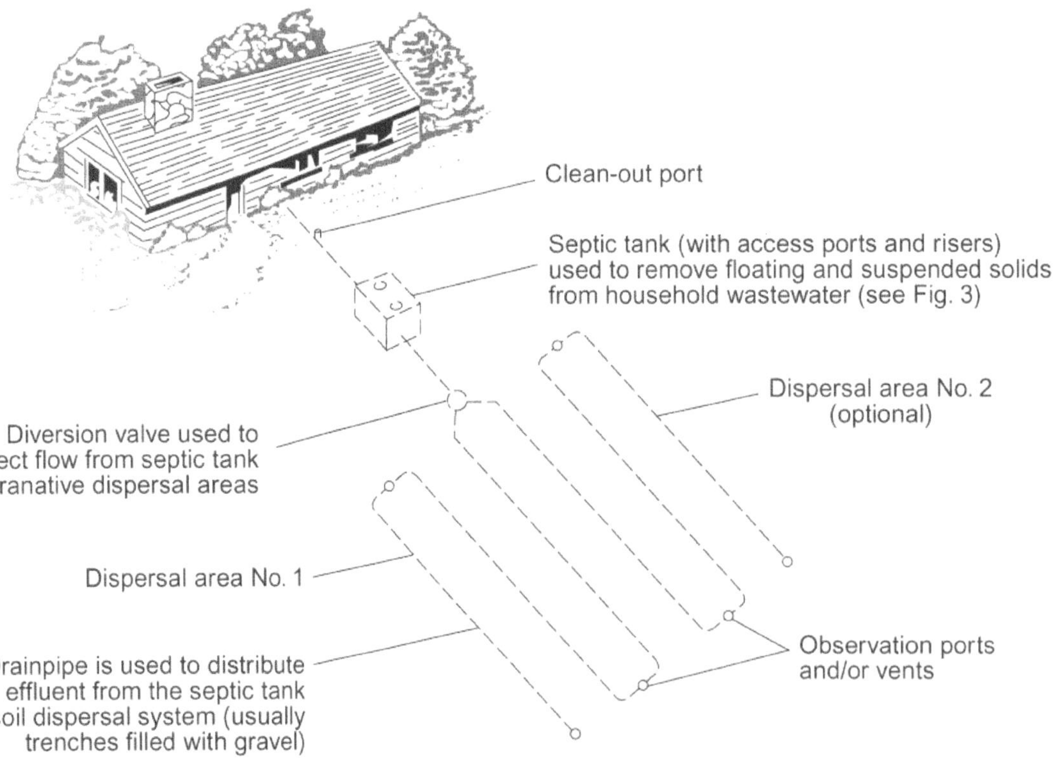

Figure 2-2. Onsite Wastewater Treatment System Consisting of a Septic Tank and Soil Dispersal System. Adapted From Tchobanoglous, G. and F.L. Burton, 1991.

2.1.3 Gas Emissions from Septic Tanks

Anaerobic degradation, occurring within the sludge layer of the septic tank, results in the production of gases composed primarily of methane and carbon dioxide. When sulfate compounds are present in the influent wastewater, hydrogen sulfide and other sulfur containing gases may also be formed. Gases formed in the septic tank are evacuated typically from the system through the building drainage plumbing and vent system. Gas emissions are discussed in more detail in Section 2.4.

2.1.4 Venting Systems

Building codes require that gases formed inside the septic tank be evacuated by means of a vent system. Household plumbing vents are used generally as vents for the septic tanks. Less commonly, gases may also be vented through the leach field or through screened atmospheric pipes located inside the tank (D'Amato et al., 2008). The two main purposes of tank vents are to avoid wastewater backflow due to a vacuum created inside the house plumbing fixtures and to let toxic, odorous (e.g., hydrogen sulfide, mercaptans), and explosive (e.g., methane) gases formed during the anaerobic degradation escape and be diluted in the atmosphere (Kaplan, 1991).

Gases such as hydrogen sulfide often generate concern due to odor generation, potential human toxicity, and ignition properties (D'Amato et al., 2008; EPA-IRIS, 2009). Nevertheless, based on measurements reported by Winneberger (1984), the hydrogen sulfide concentration from vented tanks are below detection limits and, therefore, gas evacuation throughout venting systems does not represent a fire risk for the household residents. It has been shown that gases in the headspace of the tank escape through the inlet and outlet tees and eventually to the house vents (Winneberger, 1984).

2.1.5 Effluent Dispersal

In a conventional septic system, clarified effluent is discharged typically to a soil dispersal system (see Figure 2-2). The soil dispersal system receives the treated wastewater and distributes it into the soil through a perforated pipe system located in gravel filled excavated trenches (U.S. EPA, 2003). The soil operates as a biofilter, where biological, physical, and chemical processes take place. Inorganic and organic compounds may be transformed to various degrees in the soil while pathogens die off.

Operationally, problems can occur in the soil dispersal system when the application of septic tank effluent exceeds the infiltration capacity of the soil. The infiltration capacity of the soil is a function of the soil properties and characteristics of the septic tank effluent. In general, loading of dissolved organic matter supports the growth of biomass that restricts soil pore space, while the loading of particulate matter fills and blocks soil pores (Leverenz et al., 2009). Thus, a high loading of both dissolved and particulate organic matter will result in a reduced infiltration rate. In the extreme case, effluent can surface above the soil dispersal field, which is an indication that the hydraulic loading rate has exceeded the soil infiltration rate for the given loading scenario (Crites and Tchobanoglous, 1998). Proper septic tank design, along with regular monitoring and maintenance, can be used to control the discharge of constituents that will reduce the infiltration capacity. The discharge of chemical and biological constituents to groundwater is also a concern associated with onsite wastewater systems.

2.2 Septic Tank Characteristics

Septic tanks are considered simple and effective primary treatment systems; however, there are several important processes that must occur to achieve a satisfactory level of treatment. General system configurations, hydraulics, processes occurring within septic tanks, impacts of invertebrate animals, and a description of the tank operation and maintenance, with a focus on sludge accumulation and solids extraction frequency, are described in this section.

2.2.1 Process Configuration

The general configuration of a septic tank is shown on Figure. 2-3. The tank liquid capacity varies between 2.8 and 5.6 m^3 (750 and 1500 gal) when used for a single house (Tchobanoglous and Schroeder, 1985), with specific tank sizing based on parameters such as the number of inhabitants, the home maximum occupant capacity, and use of water-saving fixtures (U.S. EPA, 2000). Septic tanks are constructed typically of concrete or fiberglass, but can also be made of polyethylene (Crites and Tchobanoglous, 1998; U.S. EPA, 2000).

As shown on Figure 2-3, most modern septic tanks have access ports or risers at grade for inspection and cleaning activities. Riser lids should be secured or locked to prevent unauthorized access. Septic tanks are generally buried in the ground and must be watertight and structurally sound to prevent leakage and eventual failure (U.S. EPA, 2000). Other components are the inlet tee, effluent filter (optional), and outlet tee, which are designed to retain solids in the tank while allowing the clarified water and gases to move through the tank. Ribbed risers are sometimes not recommended in cold climates where the soil freezing can uplift and displace the risers from the septic tank.

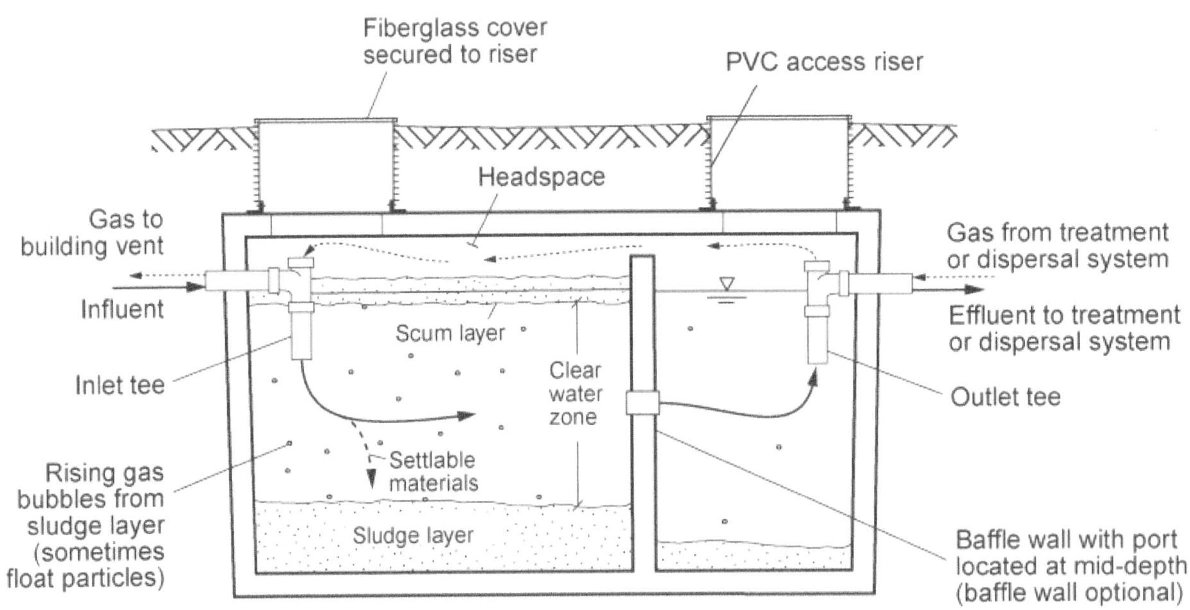

Figure 2-3. Sectional View of Dual Compartment Septic Tank.
Adapted from Crites and Tchobanoglous, 1998.

2.2.2 Tank Hydraulics

A septic tank can have one or two compartments (Crites and Tchobanoglous, 1998). The two-compartment tank is recommended by the uniform plumbing code (UPC) to aid in the retention of solids (Perkins, 1989; D'Amato et al., 2008); however, Winneberger (1984) found that a single compartment tank performed as well, with respect to effluent quality, as a two-compartment model with the same capacity. One explanation is that a single compartment tank has a larger surface area available, and therefore the settling may be more efficient (Crites and Tchobanoglous, 1998). In general, the results of the septic tank compartment findings have been ambiguous due to the lack of long term studies (Bounds, 1997). Different baffle configurations can be used to improve tank hydraulics and facilitate sludge removal.

In general, the tees control the flow of liquids and solids in the tank and act as a pathway for gases to leave the tank into the house vent system. Tees also help to avoid short-circuiting of the wastewater through the tank to the outlet and prevent mixing between the scum and the incoming liquid (Bounds, 1997; U.S. EPA, 2002). The invert elevation of the outlet tee is generally located 2 to 3 inch below the invert elevation of the inlet tee to keep the inlet pipe above the water level. The rising leg of the inlet tee should extend up in length at least 6 inch over the liquid level to prevent the scum layer from clogging the inlet (U.S. EPA, 2000).

The inlet tee (see Figure 2-3) allows the incoming wastewater to be introduced without disturbing the clarification process that is taking place inside the tank (Winneberger, 1984; Perkins, 1989). Further, the tee minimizes short-circuiting (U.S. EPA, 2002) and enhances sedimentation and detention time (Bounds, 1997). The outlet tee prevents scum and floating sludge from passing through and clogging the soil dispersal system (Winneberger 1984; Perkins, 1989).

The hydraulic retention time (HRT) in a septic tank is directly related to the tank geometry. Of tanks with similar volumes, shallow tanks with greater surface area have lower overflow rates and as a result more efficient capture of solids, while deep tanks can store more settled solids (U.S. EPA, 2000; D'Amato et al., 2008). The typical theoretical HRT for septic tanks varies from 24 to 72 hr (D'Amato et al., 2008); higher values of HRT ranging from 60 to 80 hr have also been reported (Walker and Driftmier, 1929). Other studies recommend values ranging from 6 to 24 hr (Winneberger, 1984; Bounds, 1997). A summary of HRT recommendations is presented in Table 2-1.

Table 2-1. Recommended Hydraulic Retention Time in Septic Tanks.

Range (hr)	Reference
6 – 24	Winneberger (1984)
	Bounds (1994)
24 – 72	D'Amato et al. (2008)
60 – 72	Walker and Driftmier (1929)

The actual HRT depends on the geometry, depth, number of compartments, solids volume, and inlet and outlet designs of the septic tank and thus varies greatly from tank to tank (U.S. EPA, 2002). When hydraulic overloading occurs, the usual retention time of the septic tank might not be long enough to allow for effective settling, resulting in solids flow through the tank outlet and obstruction of the effluent dispersal system or downstream process.

2.2.3 General Conversion Processes Occurring in Septic Tanks

In a simplistic view, the septic tank operates as a settling basin allowing the influent wastewater particles to settle to the bottom and form a solids layer known as sludge (U.S. EPA, 2003). Greases, oils, and other buoyant particles rise to the water surface and form a scum layer composed of accumulated floating materials (Crites and Tchobanoglous, 1998; U.S. EPA, 2005). As depicted on Figure 2-3, a septic tank would generally have three characteristic layers: scum at the top, a clear zone in the middle, and the sludge layer at the bottom. The clarified water flows between the scum and sludge layers and leaves the septic tank for further treatment (if present) or soil dispersal.

The major mechanism of oxygen demand (OD) removal from septic tank effluent results from the settling of suspended solids from the influent wastewater to the bottom of the tank. While the liquid remains in the tank one or two days (see Table 2-1), the settled solids remain in the tank and undergo anaerobic decomposition over a long period of time, for example 5 to 15 yr (Rittman and McCarty, 2001). The biochemical oxygen demand (BOD) of the septic tank effluent is typically 25-50% less than the BOD of the influent (Crites and Tchobanoglous, 1998; U.S. EPA, 2005).

The proper operation of a septic tank is based, in part, on the development of a facultative and anaerobic environment in which microorganisms perform complex biochemical reactions. The bottom portion of the septic tank behaves, for the most part, as an anaerobic reactor; however, at or near the water surface, the presence of oxygen results in facultative as well as aerobic reactions (D'Amato et al., 2008). Wastewater characteristics such as temperature, organic loading, hydraulic loading, detergents, chemicals, and cleaning products can encourage or inhibit microbial development (Bounds, 1997). For example, it has been reported that gas production and digestion activity in sludge is temporarily inhibited by discharge of water softener brine to septic tanks (Seabloom et al., 2005).

Organic material entering the sludge layer of the septic tank undergoes hydrolysis, where bacteria partially transform complex organic molecules into simple molecules, a process also known as liquefaction (Crites and Tchobanoglous, 1998; D'Amato et al., 2008). Simple organic molecules are then converted to short chain length volatile fatty acids (VFAs) by acid-forming bacteria in a process known as acidogenesis, resulting in a decrease in the water pH. At this point, methanogens begin to convert the VFAs into carbon dioxide and methane in a process called methanogenesis (Crites and Tchobanoglous, 1998; D'Amato et al., 2008). A more detailed description of the methanogenesis process is presented in Section 2.3.

Methanogenesis occurs spontaneously in septic tanks, but the conditions that determine the start of anaerobic digestion have not been elucidated fully. It has been reported that, based on observations of gas production, the methanogenesis activity inside community septic tanks is readily established (Kinnicutt et al., 1910) perhaps due to the high organic loading rate (OLR), but may not reach a peak until two years of operation in a septic tank for an individual home (Philip et al., 1993), which are usually operated at a lower OLR. Weibel et al. (1949) determined that the use of a seed inoculum (anaerobic sludge from an Imhoff tank) added at startup at a rate of about 23 L/capita·d resulted in the rapid development of anaerobic digestion. It should be noted that tank seeding is not practiced commonly.

In studies on septic tanks carried out by Weibel et al. (1955), it was reported that for a small capacity tank started in the winter, a five month lag phase was observed. The lag phase was characterized by low pH and offensive odors, which are associated with hydrolysis and acidification. Methanogenesis began to occur gradually as the temperature increased during summer. The onset of methanogenesis was characterized by increased pH, reduced effluent volatile suspended solids, and a reduction in odor. For the same tank, cleaned out and restarted in the spring, there was a shorter lag phase, which was not accompanied by low pH or offensive odors. Weibel et al. (1955) suggested that the effects of process startup may be less extreme in larger tanks and that a seed inoculum be considered for starting tanks in the winter to avoid odor generation.

Methane can also be consumed by methanotrophic bacteria under aerobic conditions, converting the methane to carbon dioxide. Methane consumption has been observed in some anaerobic/aerobic environments, but has not been measured in septic tank systems. It is reasonable to assume that methanotrophic activity could exist in septic tanks at the air-water or air-scum interface where oxygen and methane are present together (Knowles, 1993).

2.2.4 Invertebrate Communities in Septic Tanks

Invertebrate communities are commonly found within septic tanks; however the role of the individual species on degradation processes occurring in the tank is not clear. Dunbar (1907) observed that thick scum layers were composed of condensed plant remains, paper, hair, and other residual matter. According to Dunbar, this condensed material was transformed by fungi action (e.g., *Peziza omphalodes*) into a mass, where a variety of invertebrates such as worms, earthworms, and larvae of some insects (in particular, Psychoda phaloenoides) were encountered. Nevertheless, a clear relation between the action of invertebrate animals and the scum thickness was not entirely demonstrated.

In a study carried out at Novato, California, it was found that Diptera, Collembola, Acarina, Nematoda, Isopoda, and Oligochaeta are the dominant species living within the septic tanks (Dale, 1982). Dale reported that the thickness of the scum layer was directly related to the number of organisms present. A larger number of organisms corresponded to a thicker layer of scum.

The visual characteristics of the scum depend, in part, on the type of invertebrate animal species present. For example, Dale observed that scum layers had a humus look when *Lumbricidae* were present. Winneberger (1984) also reported that the scum upper layers were usually dark brown to black and had the appearance of "crumbly earth" or humus, and were frequently populated by earthworms. Thus, it appears that the configuration of the scum layers involves more than buoyant solids rising to the liquid surface; some invertebrate animals may also participate in its development, degradation, and thickness.

2.2.5 Operation and Maintenance

As stated previously, incoming solids accumulate in the tank forming a sludge and scum layer. Sludge and scum accumulation depend on several factors including tank design, the use of garbage grinders, user diet, season of the year, and temperature. These factors and their influence on the sludge accumulation are summarized in Table 2-2. Both scum and sludge layers will increase in thickness with daily solids additions and eventually the removal of all of the accumulated solids from the tank is necessary to avoid malfunctioning of the system.

Based on a number of studies, it has been found that the sludge and scum accumulation rates are highly influenced by the temperature and season of the year. Walker and Driftmier (1929) reported that the thickest scum measurements were obtained during early summer when increasing temperatures were recorded. Walker and Driftmier (1929) also reported an increase in sludge accumulation during the winter months, when bacterial degradation is hindered by the lower temperatures. These observations also support the concept of the 'spring boil', where high rates of gas bubble release are observed (see also Section 2.4.1). During the winter months, when the temperature in the septic tank cools, the rate of digestion slows and solids begin to accumulate, increasing the thickness of the sludge layer. Microbial activity in the sludge layer increases when the tank contents warm during the spring and early summer, resulting in an increase in gas production. Also, the solubility of dissolved gases decreases as the contents of the tank become warmer, also increasing the release of gases. The gas bubbles may rise to the surface individually, but also float solids to the surface. The buoyant solids then become part of the scum layer, thus increasing the thickness of the scum layer, or are discharged with the effluent.

Table 2-2. Summary of Factors Affecting the Accumulation of Sludge and Scum in Septic Tanks.

Factor	Summary	Reference
Food waste grinder	A total increase of 77% (210% increase in scum and 31 % increase in sludge) in the amount of sludge and scum from households using food waste grinders.	Weibel et al. (1955)
	The use of food waste grinders enhanced the buildup of scum by 34%, while sludge accumulation increased by 2%.	Bounds (1997)
	Homes with food waste grinders had an increase of total suspended solids in the septic tank of 25 - 40 % compared to houses without them.	U.S. EPA (2002)
	Houses with food waste grinders have been reported to accumulate sludge and scum at a higher rate.	D'Amato et al. (2008)
House occupant diet	It was observed during routine inspections that tanks from household occupants with preferences for vegetarian diets developed thin or no scum layers.	Ball (2009)
	It was observed that the tank for a vegetarian family did not develop a scum layer compared to households with a conventional meat-based diet.	Winneberger (1984)
Season and temperature	Scum thickness increased during highest recorded temperatures of the year (early summer). During winter, when bacterial activity decreased, a thick layer of sludge was noticed.	Walker and Driftmier (1929)
	In cold months (winter season) solids built up rapidly in the bottom of the tank.	Winneberger (1984)
	Solids accumulation rate increased during the winter months.	D'Amato et al. (2008)

Solids accumulation rates are also related to the maintenance of the septic system. Based on these rates, the cleaning frequency of the tank can be determined. Other factors that influence the solids accumulation and pumping frequency are the size of the tank, specific design, number of people in the home, water usage, and household water fixtures (Weibel et al., 1955; Bounds, 1997; U.S. EPA, 2002; D'Amato et al., 2008).

Data on the average rates of sludge and scum accumulation in septic tanks from various studies are summarized on Figure 2-4 (Bounds, 1997; Crites and Tchobanoglous, 1998). The different sludge accumulation values in septic tanks reported in the literature are shown on Figure. 2-4.

Reported sludge pumping intervals differ from study to study, and sometimes are not even reported (U.S. EPA, 2002). Other pumping intervals recommended are based on the percentage of solids accumulation inside the tank. A summary of sludge extraction periods recommended in various studies is presented in Table 2-3.

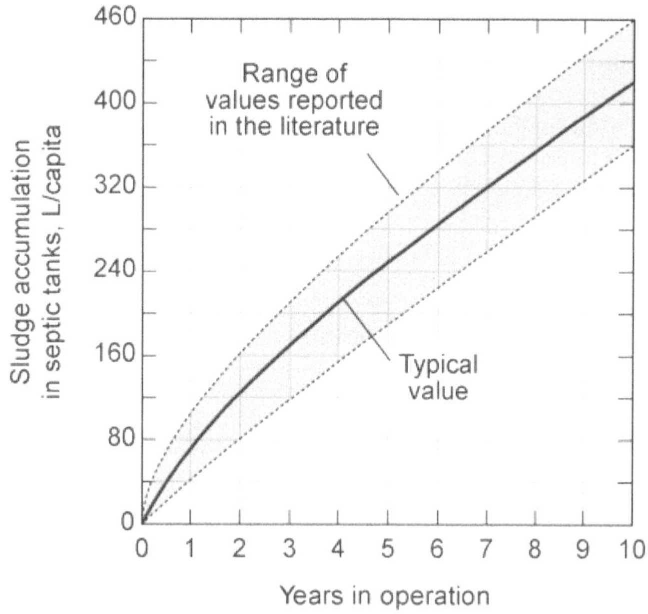

Figure 2-4. Sludge and Scum Accumulation Rates from Different Studies.
Adapted from Crites and Tchobanoglous, 1998.

As reported in Table 2-3, there is little agreement on the appropriate sludge withdrawal frequency; thus, the reported intervals can only be considered to be a guideline for pumping protocols. In some studies, equations have been given that can be used to predict the septage (i.e., entire tank contents) pump intervals based on different variables such as loading, tank size, house occupants, and use of a garbage disposal (Weibel et al., 1955; Bounds, 1997). However, regular inspection and an improved understanding of the bacterial activity in the different layers of the tank and the carbon degradation pathways are needed.

Although septic tanks are presumed to require little maintenance, periodic inspection of the scum and sludge layers, watertightness, and structural soundness are recommended to avoid environmental damage and/or health risks (Crites and Tchobanoglous, 1998; U.S. EPA, 2002). Operation and inspection guidelines have been published for septic tank owners to encourage routine inspections to protect the system from failure (U.S. EPA, 2005).

Table 2-3. Summary of Septic Tank Pumping Guidelines Reported in the Literature.

Guideline reported	Reference
Scum layer within 3 inch of the outlet elevation or the sludge layer within 6 inch of the outlet elevation	U.S. PHS (1957)
Sludge plus scum greater than ½ to ⅔ of tank depth	U.S. EPA (2002)
Sludge plus scum equal to ⅓ of tank volume	U.S. EPA (2002)
Sludge plus scum equal ¼ of the tank volume	GDPUD (2009)[c]
Every 2 years	U.S. EPA (2002)
Every 3 to 5 years[a]	U.S. EPA (2002)[b]
Every 5 years or more	Phillip et al. (1993)
Every 10 years	SLOC (2008)
Floating sludge and scum flowing through the outlet tee	U.S. EPA (2002)

[a] Range applied when regular inspections have not been done during those years.
[b] U.S. EPA recommends monitoring of the sludge and scum layers every two years and then using the natural accumulation rate for that area or that tank as the guide to determination of pumping frequency.
[c] Specific example of regulatory standards applied to the community based on site characteristics.

2.3 Anaerobic Processes

Because sludge accumulating in the bottom of a septic tank undergoes anaerobic decomposition, it is important to review the fundamental principles underlying anaerobic processes to develop a more comprehensive view of the chemical and biological reactions occurring inside a septic tank. The essential reactions that occur during anaerobic decomposition, the key operational parameters, and toxic conditions that affect the process are presented and discussed in this section

2.3.1 Anaerobic Oxidation

The anaerobic oxidation process can be described as a two-stage process (see Figure 2-5); the first stage is identified as *waste conversion (acetogenesis, acidogenesis)*, in which complex organics are first hydrolyzed and then fermented into simple organic compounds (e.g., hydrogen and carbon dioxide) and VFAs (e.g., acetate) by facultative bacteria known as acetogens and acidogens (McCarty, 1964; Speece, 1996; Bitton 2005). After the organic matter has been converted to simpler compounds, *waste stabilization (methanogenesis)* takes place, where the acids are synthesized by methanogens into methane and carbon dioxide (McCarty, 1964). It should be noted that in some references anaerobic oxidation is considered to be a three-stage process in which the hydrolysis of the organic material to simple acids is considered to be a separate step (see Figure 2-6).

A key factor in the anaerobic oxidation process is the balance between the microorganisms responsible for each step. When the system is in equilibrium, the methanogens transform the acids at the same rate that acids are formed (McCarty, 1964). Therefore, when high acid concentrations are found, it is an indication that the acid forming bacteria and the methanogens are not in balance.

The two major mechanisms of methane formation are the breakdown of acetic acid, which is the most prevalent volatile acid produced in the fermentation of carbohydrates, proteins, and fats, and the reduction of carbon dioxide (McCarty, 1964; Bitton, 2005). The chemical reactions of methane formation are as follow:

1. Utilization of acetic acid:

$$CH_3COOH \rightarrow CH_4 + CO_2 \quad \text{(Eq. 2-1)}$$

2. Reduction of carbon dioxide:

$$CO_2 + 8H \rightarrow CH_4 + 2H_2O \quad \text{(Eq. 2-2)}$$

Growth and acid utilization rates of methane formers are slow, and are usually limiting factors in anaerobic treatment (Speece, 1996, Duncan and Horan, 2003). Methanogens are known to be different from the typical bacteria and are classified in a separate kingdom, the Archaea (Duncan and Horan, 2003). The methane forming microorganisms are strict anaerobes and even small amounts of oxygen can be toxic. Methanogens are also sensitive to any environmental change including temperature, organic loading, waste composition, and other factors (McCarty, 1964). The microorganisms involved in the anaerobic process need sufficient concentrations of nutrients to operate properly. Nitrogen and phosphorus comprise about 11% and 2% of the dry weight of biological solids, respectively (McCarty, 1964).

The methanogenic organisms are restricted in the number of reactions and substrates they can utilize. Moreover, according to their substrate specificity, methanogens are classified in two groups, 1) the *Acetoclastic Metanogens*, which in general are able to utilize acetate (e.g., *Methanosaeta* spp. and *Methanosarchina* spp.) and in some cases are capable of using

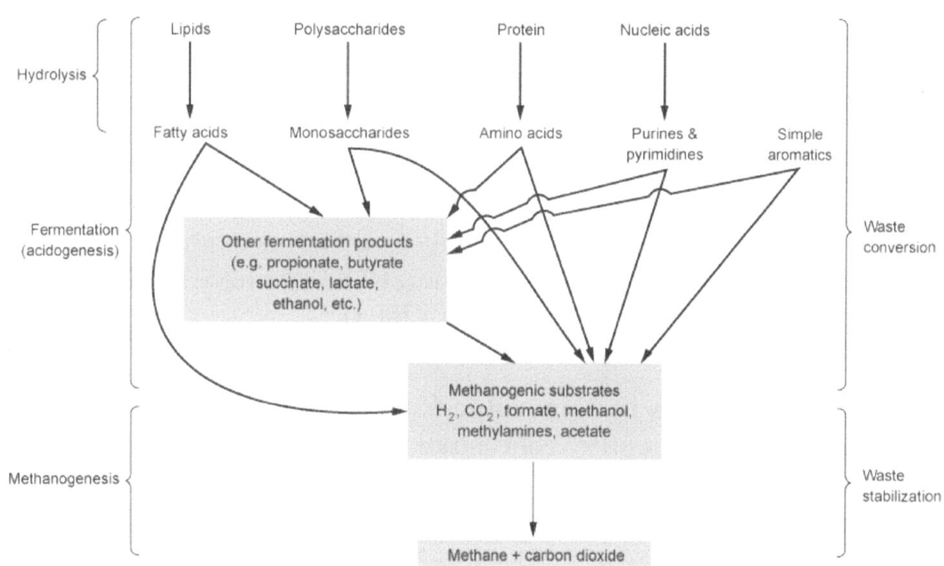

Figure 2-5. The Stages of Anaerobic Decomposition, Waste Conversion Followed by Waste Stabilization. Adapted from McCarty, 1964 and Tchobanoglous et al., 2003.

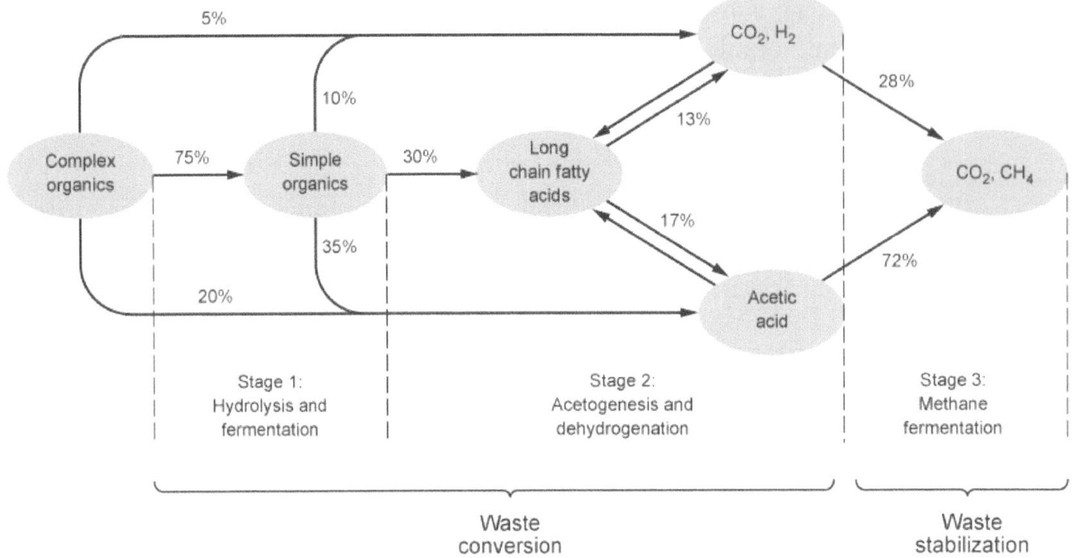

Figure 2-6. The Intermediate Steps of Anaerobic Reactions, with Percentages Based on COD Conversion.
Adapted from Speece, 1996 and Tchobanoglous et al., 2003.

methanol and methylamines (e.g., *Methanosarcina* spp.), and 2) the *hydrogen-utilizing methanogens* that reduce carbon dioxide, formate, methanol, and methylamines, using the hydrogen produced previously during the hydrolysis and acid formation processes (Duncan and Horan, 2003). In anaerobic digesters, 70% of the methane gas is originated from acetate reduction and 30% is attributed to the substrates reduced by the hydrogen-utilizing methanogens (Duncan and Horan, 2003).

A COD balance can be used to estimate the theoretical methane production during anaerobic fermentation. The COD of the methane produced during anaerobic decomposition of organic matter is approximately equal to the COD of the converted organic matter (Tchobanoglous et al., 2003). The oxygen demand of methane is determined as follows (McCarty, 1964):

$$CH_4 + 2O_2 \rightarrow CO_2 + H_2O \qquad \text{(Eq. 2-3)}$$

From Eq. 2-3, it can be derived that each mol of methane is oxidized with two mol of oxygen. Therefore, 0.35 L of methane is equal to one g COD stabilized (5.62 ft^3 CH_4 / lb COD) (McCarty, 1964; Speece, 1996; Tchobanoglous et al., 2003).

2.3.2 Operational Parameters in Anaerobic Reactors

Parameters such as the appropriate concentration of nutrients, pH, redox potential, alkalinity, volatile suspended solids loading, temperature, and solids retention time are important for successful anaerobic digestion.

In anaerobic reactors, the recommended nutrient loading is 5 to 15 mg N/g COD and 0.8 to 2.5 mg P/g COD. Nutrients must be supplied in this range as a preventive measure to avoid inhibiting effects (Speece, 1996). Sulfur is also required by methanogens, but in relatively lower concentrations, for optimal growth and maximum methanogenesis activity (Speece, 1996; Tchobanoglous et al., 2003). For comparison, typical nitrogen and phosphorus concentrations in untreated wastewater from individual residences are 13.3 and 3.28 g/capita·d, respectively

(Crites and Tchobanoglus, 1998). Based on typical operation, the estimated loading of nitrogen and phosphorus in the septic tank sludge is approximately 11 and 6.2 mg/g COD, respectively (see Table 2-6).

The pH is another important factor, which should be in the range of 6.5 to 8.2, with an optimum range of about 7.0 to 7.2 (McCarty, 1964; Speece, 1996). Outside of these ranges, the efficiency of the treatment decreases rapidly, and acid conditions can dominate the reactor environment resulting in toxic conditions for methanogens (McCarty, 1964). Therefore, controlling the reactor pH at the optimum levels is essential for efficient methanogenesis.

Microbial respiration requires an electron acceptor, which can be an organic compound or some inorganic electron acceptors, as shown in Table 2-5. Some of the most common electron acceptors in decreasing order of reduction potential are nitrate, manganese, iron, sulfate (all under anoxic conditions) and finally carbon dioxide, under anaerobic conditions (Maier et al., 2009). As reported in Table 2-5, aerobic reactions using oxygen as a terminal electron acceptor provides the most energy for cell growth, while methanogenesis provides the least. Thus, on the basis of energetics, other electron acceptors must be exhausted before methanogenesis can take place. Based on the relationships shown in Table 2-5, measurements of redox potential can be used to assess the types of reactions likely to be occurring within a septic tank.

Table 2-4. Optimum Conditions for Anaerobic Treatment.
Adapted from McCarty, 1964.

Parameter	Unit	Optimum condition
Temperature	°C	30 - 38[a] 50 - 57[b]
Environment	-	Anaerobic
Biological nutrients available	-	Nitrogen, phosphorous, and sulfur
pH	unitless	7.0 to 7.2
Toxic materials	-	Total absence
Alkalinity	mg/L as $CaCO_3$	2,500 to 5,000

[a] Mesophilic conditions.
[b] Thermophilic conditions.

Table 2-5. Common Electron Acceptors, Products, and Redox Potentials.
Adapted from Maier et al., 2009.

Type of respiration	Reduction reaction electron acceptor→product	Reduction potential (V)	Difference [a,b] (V)
Aerobic	$O_2 - H_2O$	+ 0.81	- 1.28
Denitrification	NO^{3-} - N_2	+ 0.75	- 1.22
Manganese reduction	Mn^{4+} - Mn^{2+}	+ 0.55	- 1.02
Nitrate reduction	NO^{3-} - NH_4^+	+ 0.36	- 0.83
Sulfate reduction	SO_4^{2-} - HS^-, H_2S	- 0.22	- 0.25
Methanogenesis	$CO_2 - CH_4$	- 0.25	- 0.22

[a] CH_2O-CO_2 was used as electron donor in each case, with an oxidation potential equal to – 0.47 V.
[b] Reduction - oxidation potential of CH_2-CO_2.

In anaerobic reactors, the alkalinity is controlled by the bicarbonate ion concentration, associated with the production of carbon dioxide gas. Optimum levels of alkalinity in a complete-mix high-rate anaerobic digester vary from 2,500 to 5,000 mg/L as $CaCO_3$, with a minimum of 1,000 mg/L as $CaCO_3$ to provide enough buffer capacity (McCarty, 1964). When the levels of alkalinity are less than 1,000 mg/L as $CaCO_3$, the pH will drop and the methanogenesis rate will be reduced (McCarty, 1964; Speece, 1996).

Similar to anaerobic digesters, the anaerobic sludge layer in the septic tank requires enough alkalinity to buffer against pH change due to the production of volatile fatty acids (VFAs). However, the formation of VFAs depends directly on the incoming organic loading rate (OLR), with a higher OLR resulting in a higher concentration of VFAs in the septic tank. The OLR in anaerobic reactors is usually high (3.2 to 32 kg COD/m^3·d) as compared to 0.07 kg COD/m^3·d in septic tanks (see Appendix A for calculation). While high alkalinity concentrations (e.g., 2,500 to 5,000 mg/L as $CaCO_3$) are not required to buffer acidic conditions inside the septic tank, the estimated alkalinity in the septic tank sludge layer is on the order of 5,000 mg/L (calculated using septage values from U.S. EPA, 1994).

Temperature is also a relevant parameter affecting the rate of anaerobic digestion. Thermophilic temperatures are ideal in anaerobic reactors because the reaction rates increase and the process is more efficient (McCarty, 1964). Temperature also affects ionization fractions, the solubility of substrates, and iron bioavailability (Speece, 1996). In the anaerobic consortium, methanogens are more sensitive to small changes in temperature compared to acid-forming bacteria. As temperature reduces, acidogens produce VFAs faster than methanogens can convert the VFAs to methane, creating an unbalanced metabolism during the anaerobic process (Speece, 1996; Bitton, 2005). The two temperature ranges specified for anaerobic treatments are reported in Table 2-4. Nevertheless, keeping the temperature at thermophilic ranges can be challenging, especially when the incoming waste is diluted, because the methane production might not be sufficient to heat the process using gas combustion (McCarty, 1964; Speece, 1996). In anaerobic digesters, a supplemental external heat source is often used. The liquid temperature in a septic tank is basically uncontrolled and related to hot and cold water use, as well as mean annual temperature. It should be noted that low temperatures in septic tanks are compensated by long SRT, as discussed below. Additional information on temperature in septic tanks is discussed in Section 2.4.1.

Solids retention time (SRT) in anaerobic reactors is around 20 days for processes occurring at mesophilic temperatures (i.e., 30°C). The suggested SRT can be as high as 28 days at temperatures of 18°C, and as low as 10 d for processes occurring at 40°C (McCarty, 1964; Tchobanoglous et al., 2003). In a complete-mix digester (i.e., typical anaerobic digester) the SRT is the same as the HRT (Tchobanoglous et al., 2003). However, a septic tank is not a complete mix-reactor because sedimentation and solids accumulation processes are involved and result in a phase separation. It should be noted that limited intermittent mixing does occur in the sludge layer in a septic tank due to bubble formation and release.

Sludge accumulates in the septic tank for years between tank cleaning events, resulting in an extended SRT, while the design HRT in the tank is only 1 to 2 d (U.S. EPA, 2002). In a complete-mix anaerobic reactor, 60% of volatile suspended solids (VSS) are destroyed in 20 d (Tchobanoglous et al., 2003). Therefore, the VSS destruction in a septic tank is expected to be relatively high (e.g., 50%) due to the long SRT (see Appendix B).

Solids are removed from the tank periodically, usually in response to poor performance or when the volume of solids in the tank reaches a certain limit (see Table 2-3). At the time of cleaning, it is common to have the septic tank emptied fully without leaving sludge in the tank for seeding purposes (U.S. EPA, 2002). The material that is removed from the tank is known as septage. It should be noted that the solids contained in the septage range from old material accumulated since the previous tank cleaning event to material deposited immediately prior to cleaning. Thus, the SRT is approximately equal to one half of the time between tank cleaning events. A comparison of the operational parameters for a typical anaerobic digestion process and a septic tank is shown in Table 2-6.

2.3.3 Toxic Compounds in Anaerobic Reactors

There are many inorganic and organic compounds that at certain concentrations become inhibitory or toxic in anaerobic reactors. Toxic compounds can range from inorganic ions such as potassium, magnesium, or calcium, to metals such as copper, zinc, or lead (McCarty, 1964). In general, a toxic effect occurs at high concentrations. Conversely, at low concentrations, the effect of these compounds might be stimulatory and favorable to the anaerobic process (McCarty, 1964). Moreover, anaerobic conditions must be maintained in the reactor as even small amounts of oxygen inhibit methane forming microorganisms.

Earth metal salts commonly found in industrial wastes, such as sodium, potassium, calcium, or magnesium, are highly toxic, causing failure or low treatment efficiency (McCarty, 1964; Speece, 1996). It has also been reported that toxicity is related directly to the cation of the salt

Table 2-6. Typical Operating Parameters for Single-Stage Complete Mix Anaerobic Digester and Septic Tank.

Parameter	Units	Single-stage complete mix digester	Septic tank
COD	$kg/m^3 \cdot d$	3.2 - 32	0.07 – 0.106[a]
SRT	d	20	574 – 730[b,c]
HRT	d	20	1 – 2
Temperature	°C	30 - 38	7 – 30
Nitrogen	mg/g COD	5 – 15[d]	11[e]
Phosphorus	mg/g COD	0.8 – 2.5[d]	6.8[e]
Alkalinity	mg/L as $CaCO_3$	2,500 to 5,000	4,500
Mixing	-	Complete	Intermittent by bubble activity

[a] Calculated values, see Appendix A.
[b] Approximately equal to one half of the time between tank cleaning events.
[c] Cleaning event assumed every three to five years (D'Amato et al., 2008).
[d] Rittman and McCarty (2001).
[e] Based on typical raw residential wastewater values (Crites and Tchobanoglous, 1998; Höglund, 2001).

(i.e., Na^+) instead of the anion (McCarty, 1964). For example, in a septic tank study, Weibel et al. (1955) reported that a 1.2 % (12,000 mg/L) mixed salt brine representative of a water softener backwash cycle inhibited a non-acclimated anaerobic digestion process for 9.5 d. Weibel et al., (1955) also reported that digestion activity in acclimated sludge was not inhibited at representative concentrations. Inhibitory as well as stimulatory concentrations of selected compounds are shown in Table 2-7.

Table 2-7. Cation Concentrations that Cause Inhibitory and Stimulatory Effects on Anaerobic Processes[a].

Cation	Concentration, mg/L		
	Strong inhibition[b]	Moderate inhibition[c]	Stimulatory
Na^+	8,000	3,500 - 5,500	100-200
K^+	12,000	2,500 - 4,500	200-400
Ca^{2+}	8,000	2,500 - 4,500	100-200
Mg^{2+}	3,000	1,000 - 1,500	75-150

[a] Adapted from McCarty, 1964.
[b] Concentration that slows down the anaerobic treatment resulting in low efficiency.
[c] Concentration that can be accepted with some microbial acclimation.

Ammonia and ammonium ion are also toxic compounds found in anaerobic reactors produced normally during the anaerobic degradation of proteins or urea, which are present in some industrial wastes and in concentrated municipal sludge (McCarty, 1964; Speece, 1996). The ammonia concentration depends on the pH and can be present as mostly as the ammonium ion (NH_4^+) at pH values below 7.25, or mostly as dissolved ammonia gas (NH_3) at higher pH values (McCarty, 1964). Concentrations reported as inhibitory vary from 1500 to 3000 mg/L ($NH_3 + NH_4^+$) and completely toxic at concentrations above 3,000 mg/L (Rittman and McCarty, 1964).

High sulfate concentrations can be problematic during anaerobic digestion processes because sulfate reducing bacteria (SRB) compete with methanogens for substrate (i.e., acetate) within the reactor. The activity of SRB can inhibit methanogens and increase the concentration of hydrogen sulfide (H_2S), which is toxic to the majority of the biomass. Hydrogen sulfide is a corrosive gas and its removal from the biogas is expensive (Winfrey and Zeikus, 1977; Schönheit et al., 1982; Isa et al., 1986; Parkin, 1990; Speece, 1996). Ranges of sulfate inhibitory concentrations reported in the literature are presented in Table 2-8.

Table 2-8. Inhibitory Sulfate Concentrations for Anaerobic Processes Reported in the Literature.

Sulfate concentration, mg/L	Reference
>100[a]	Winfrey and Zeikus, 1977
>200[b]	Patel et al., 1978
>50[c]	Parkin and Speece, 1982
>800[d]	Parkin and Speece, 1982
>145[e]	Parkin et al., 1990

[a] Analysis performed in freshwater sediments.
[b] Pure cultures of methanogens were growth in synthetic media.
[c] Value found in an unacclimated batch digester.
[d] Value found in a submerged anaerobic filter.
[e] Value found in an anaerobic chemostats for acetate systems.

In some cases, sulfate can also be favorable for anaerobic treatment. When sulfate is reduced to sulfide by microbial action, it can combine with and precipitate metals such as copper, nickel, or zinc as non-toxic materials, resulting in an effective approach to control metal toxicity (McCarty, 1964; Isa et al., 1986). McCarty (1964) reported that concentrations below 200 mg/L are not toxic to anaerobic digestion reactors. In septic tanks, the sulfide produced during anaerobic degradation follows common metal precipitation pathways, forming insoluble metallic sulfides (Crites and Tchobanoglous, 1998) as depicted on Figure 2-7. It should be noted that sulfate reduction in septic tanks is typically incomplete; thus, sulfate is commonly detected in septic tank effluent. The increased sulfate concentration of wastewater in septic tanks resulting from water usage ranges from 30 to 60 mg/L (Crites and Tchobanoglous, 1998).

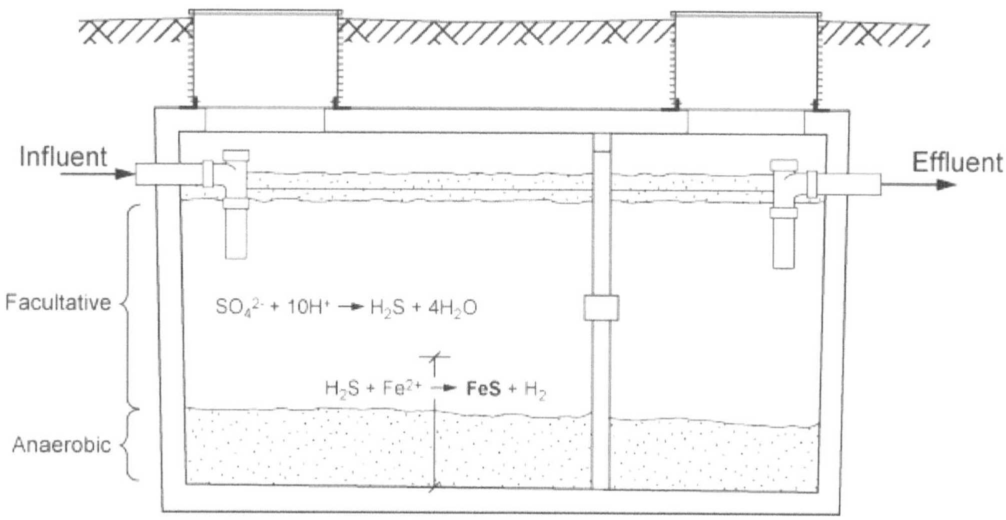

Figure 2-7. Sulfate Reaction with Metals in a Septic Tank.

Parkin et al. (1990), reported that sulfide is toxic in a range between 50 to 430 mg/L, with the toxic effects noticeable at the lower range. Moreover, Parkin et al. (1990) found that irreversible failure in anaerobic chemostats was noted at 62 mg/L for acetate systems and 60 mg/L for propionate systems. Conversely, Maillacheruvu and Parkin (1996) concluded that despite the sulfide toxicity, methanogenesis can still occur even in the presence of H_2S under specific conditions, such as high pH. Thus, it seems that sulfide dynamics and its effects on methanogenesis depend on microbial reduction, combination with metals, and pH levels within the anaerobic reactor. It is estimated that 1 mg/L of a sulfide salt such as sodium sulfide (Na_2S) is enough to precipitate soluble metals (McCarty, 1964). The weak sulfide complexes formed (i.e., FeS, highlighted on Figure 2-7) are not harmful to the microorganisms present in the sludge.

Finally, organic compounds such as alcohols and fatty acids can be toxic when fed to anaerobic reactors at high concentrations (McCarty, 1964). This situation is common when industrial wastewater is being treated; however, when the organic material is fed continuously, the anaerobic reactor acclimates and is able to handle the organic compound degradation (McCarty, 1964). Septic tanks used for the treatment of domestic wastewater are typically not subject to loading with concentrations of organic compounds sufficient to cause toxicity. In addition, the anaerobic reactions occurring in the sludge layer are buffered from toxicity associated with the liquid flowing through the tank.

2.4 Gas Emissions from Septic Systems

Even in the earliest literature on septic tanks, gas emissions have been a topic of interest. The formation of gases such as methane and carbon dioxide inside the tank are due to complex biological reactions, which are influenced by different factors such as temperature, loading, and season of the year. The focus of this section is on the factors affecting gas production, different techniques used to measure gas fluxes, and methane estimates from several sources that have reported septic tank gas emissions.

2.4.1 Gas Formation and Temperature Influence in Septic Tanks

The temperature inside a septic tank depends on the water use activities in the house and follows seasonal temperature changes according to geographic location. As shown on Figure 2-8a, in tanks located in the San Francisco Bay area, which has a temperate climate and little seasonal variation, the temperature follows the ambient temperature patterns, varying about 6 - 8°C throughout the year. However, as shown on Figure 2-8b and 2-8c for Quebec (Canada) and Kansas (U.S.), the temperature in septic tanks in more extreme climates is subject to higher seasonal variation than San Francisco.

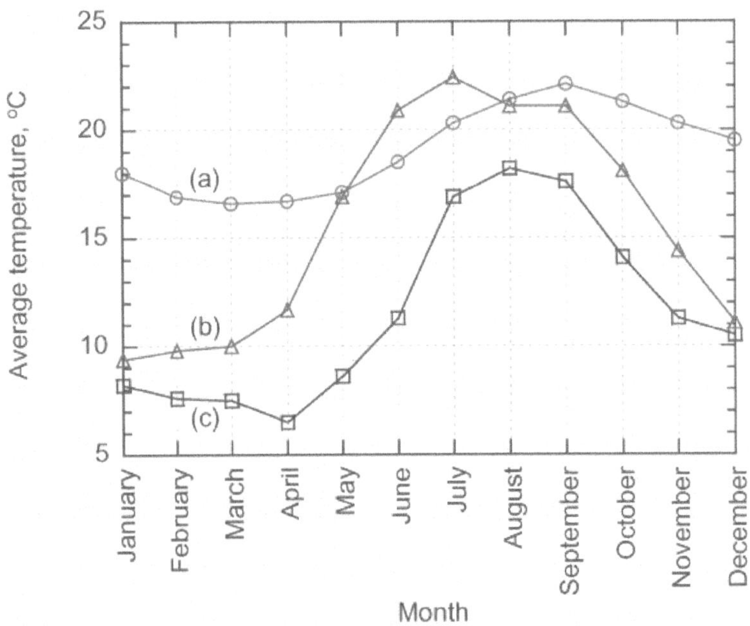

Figure 2-8. Average Monthly Temperature in Septic Tanks located in (a) San Francisco Bay Area, U.S. (Adapted from Winneberger, 1984); (b) Quebec, Canada (Adapted from Roy and Dubé, 1994) and (c) Kansas, U.S. (Adapted from Walker and Driftmier, 1929).

The gases formed during the anaerobic digestion process occurring within the sludge layer in the septic tank rise to the liquid surface or to the scum layer (if present). Settled solids accumulated on the bottom of the tank to which gas bubbles attach eventually become buoyant and rise to the surface to become part of the scum layer (Walker and Driftmier, 1929; Perking, 1989; Crites and Tchobanoglous, 1998).

The rate of gas formation inside the tank is related to temperature (Walker and Driftmier, 1929; Winneberger, 1984; D'Amato et al., 2008). Winneberger (1984) pointed out that septic tanks developed a temperature gradient from top to bottom. In the study, it was determined that warmer temperatures were located in the bottom and colder temperatures were found on the top. Based on the analysis and observations, temperature variation was correlated with hot water use in the house. In addition to the septic tank inner thermal stratification, seasonal temperature variation has also been reported (Walker and Driftmier, 1929; Winneberger, 1984; D'Amato et al., 2008), as shown previously on Figure 2-8. During cold months (winter), the rate of solids decomposition is reduced and the amount of solids build up in the bottom of the tank increases.

Conversely, in warmer months (spring) the degradation rate increases due to the elevated temperature in the accumulated solids. A sudden increase in the rate of anaerobic activity can result in a condition known as the spring turnover or spring boil (Winneberger, 1984; D'Amato et al., 2008). The increased gas production and the change in the solubility of the dissolved gases during the spring turnover results in a decrease in the solids removal efficiency due to the resuspension and discharge of settled solids. The gases also disturb the incoming solids and therefore inhibit their ability to settle (D'Amato et al., 2008).

2.4.2 Gas Collection Techniques and Chamber Systems

Theoretical estimates of gas fluxes from septic systems can be determined from the organic loading of the system or by models developed for this purpose (e.g., IPCC, 2006). Direct techniques to measure gas fluxes from septic tanks have rarely been reported in the literature (Walker and Driftmier, 1929; Winneberger, 1984). However, devices designed to measure gas fluxes from environmental systems (e.g., anaerobic ponds, wetlands, and agricultural soils) have been in use for a number of years.

Winneberger (1984) collected gas samples from septic tanks using submerged inverted bowls, which were placed inside seven tanks for several days. Subsequently, gas samples were taken and analyzed for methane, carbon dioxide, hydrogen sulfide, and other air components using standard techniques such as mass spectrometry and iodometry. One potential issue with this study is the long contact time between the collected gases and the liquid, potentially allowing for some of the gases to partition into the liquid following Henry's Law, and thus impact the relative concentration of the gases analyzed.

Measurements of gases using floating platforms with gas-collection domes or chambers have been used to determine methane fluxes in anaerobic lagoons and wetlands (Moore and Roulet, 1991; DeSutter and Ham, 2005; Aneja et al., 2006). An example of a flux chamber design to measure gas fluxes from soil-plant systems is depicted on Figure 2-9.

The method used in each study depends on the particular conditions of the systems analyzed and the research objectives. While the gas sampling methodology is highly specific to the particular system under assessment, the chamber configuration and design have certain parameters in common. Typically, non-reactive materials are used to fabricate the body of the chamber, with stainless steel and PVC being used most frequently. Teflon tubing is recommended for vents and sampling lines to minimize chemical reaction and temperature rise. Moreover, the incorporation of a small fan to mix the headspace of the chamber is seen in almost all the flux measurement devices. For those devices located in outdoor environments, an insulated cover is used to diminish temperature alteration inside the chamber headspace (Moore and Roulet, 1991; Hutchinson and Livingston, 1993; USDA-ARS, 2003; DeSutter and Ham, 2005; Aneja et al., 2006).

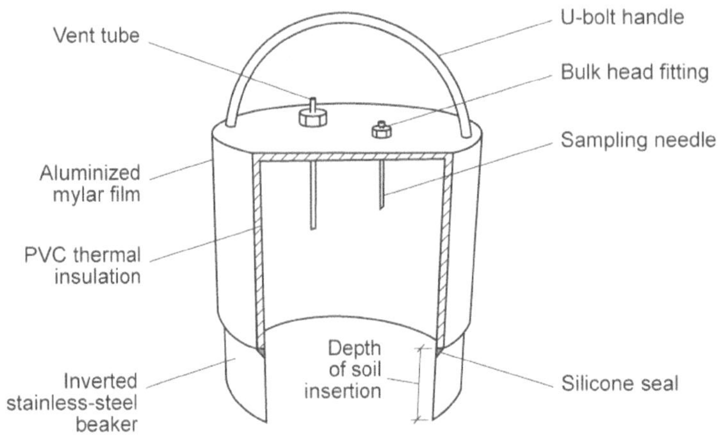

Figure 2-9. Flux Chamber Designed by the USDA-ARS GRACEnet to Measure Gas Fluxes from Soil Systems. Note the Venting Tube and PVC Materials are Common Features in Flux Chambers.
Adapted from Chamber-Based Trace Gas Flux Measurement Protocol. USDA-ARS GRACEnet, 2003.

Moore and Roulet (1991) used both open (dynamic, forced flow-through air circulation) and closed (static, no forced air exchange) flux chambers; methane fluxes were determined by means of gas chromatography (GC). From a comparison between the two types of chambers it was found that the methane fluxes may be underestimated using the static chamber by 20%. Nevertheless, the researchers suggested that statistically, the difference in performance is relatively small. The static chambers were recommended for measuring regional estimates of methane fluxes because they can be economical and low-maintenance compared with the dynamic chambers.

Hutchison and Livingston (1993) discussed that both types of chamber systems (dynamic and static) are subject to bias from physical and biological factors during the measurement process; and suggested some corrective actions according to the source of the bias. For example, to minimize pressure effects, Hutchison and Livingston (1993) recommended the use of an open tube to evacuate the air from the chamber when it is first installed. According to Hutchinson and Livingston (1993), there is not a preference to use one chamber system or the other. However, it was noted that the chamber used should be adapted to the particular environment and sampling conditions, taking into account factors such as temperature, season, atmospheric pressure, soil type, irrigation frequency, and wind.

Based on comparisons between open and closed flux chambers and spectroscopic techniques such as fourier transform infrared (FTIR) spectroscopy and tunable diode laser spectroscopy (TDLSS) to measure nitrous oxide fluxes from two soil-plant systems, it was concluded that there was a good agreement and no bias in the data was obtained with the different methodologies (Christensen et al., 1996).

2.4.3 Results from Previous Gas Measurement Studies

Gas fluxes from septic tanks have rarely been measured. Winneberger (1984) studied seven septic tanks for a ten month period. Gas samples were collected and analyzed for carbon dioxide, hydrogen sulfide, and methane, as well as other atmospheric components such as argon, oxygen, nitrogen, and hydrogen. Winneberger (1984) determined that gas measurements of carbon dioxide and hydrogen sulfide varied greatly from tank to tank. However, it was reported that methane values were more consistent, ranging from 66.3 to 88.7% by volume, with a mean of 72.9%. The highest methane value reported was in the hottest month, therefore, it is possible that the amount of gas increased due to the enhanced sludge degradation. Some important observations reported in the Winneberger (1984) study were:

- Gases formed by the sludge degradation over the tank bottom were uniformly produced.
- Gases filled up the headspace of the chamber collector apparatus in two days.
- The gas flux estimated per capita in one tank was 28 L/capita·d.

In a study conducted by Philip et al. (1993), an attempt was made to measure methane production from 50 septic tanks during a three yr period in Southern France. To analyze the methane production potential from the sludge, samples were withdrawn and placed at 20°C for three weeks. In this case, the gases were not collected directly from the septic tank by means of flux chambers or similar devices. During the third year of the study, methane production was found to have increased by a factor of five compared to the first two years. The authors concluded that the degradation of VFA and hydrolysed compounds increases after the second year due to the decline of COD. Unfortunately, the study does not go into detail on the causes of the change in gas production and no data are provided on individual septic tank gas fluxes.

Globally, it is believed that nitrous oxide (N_2O) emissions also come from various types of solid waste and wastewater management practices, including pit latrines, composting toilets, septic tanks, and engineered systems, including activated sludge, trickling filters, anaerobic or facultative lagoons (Bogner et al., 2007). These N_2O emission rates are expected to be reduced through implementation of mitigation technologies such as landfill gas recovery, improved landfill practices, engineered wastewater management, controlled composting and expanded sanitation coverage in countries under the Kyoto Protocol, especially in Asia, Africa, South America and the Caribbean (Bogner, 2007). Moreover, in an N_2O emission study developed by U.S. EPA (1999) on septic tanks and latrines in developing countries around the world, it was concluded that nitrous oxide contributions from these types of systems are probably not a significant source.

In the United Stated, a majority of nitrous oxide (N_2O) emissions are generated as a result of agricultural practices (U.S. EPA, 2009). While considered to be a relatively minor source, N_2O emissions are also generated in wastewater treatment as an intermediate product during nitrification and denitrification processes. Nitrous oxide formation is promoted by conditions of reduced aeration, high moisture, and abundant nitrogen in the form of urea, ammonia, or proteins (Bogner et al., 2007; U.S. EPA, 2009). Little specific data on emissions from septic systems have been reported in the literature, however, it is expected that nitrification and denitrification of septic tank effluent in soil-based dispersal systems could potentially result in some N_2O emissions.

2.4.4 Estimates of Methane Production

To address the lack of studies on direct measurements of GHG emissions from septic tanks, it is necessary to calculate emission values that might represent a baseline for comparison when empirical values are obtained. Estimates of methane emissions can be developed based on an organic loading approach, where it is assumed that in the United States, one person discharges 200 g/d of COD (Crites and Tchobanoglous, 1998) and that 60% of the influent COD is reduced due to settling and anaerobic digestion inside the septic tank. Based on this calculation approach, the theoretical methane production is 11.0 g CH_4/capita·d (see Appendix B for calculations). It should be noted that this calculation approach accounts for the COD that is eventually removed from the system as septage.

Kinnicutt et al. (1910) reported methane measurements from community septic tanks in Lawrence, MA in the early 1900's. Studies from Kinnicutt et al. (1910) describe the significant relation that exists between the methane emissions from septic tanks and the temperature. As shown on Figure 2-10, values around 8-12°C have corresponding methane emission rates in the range of 0.1 to 2 g CH_4/capita·d. Alternately, temperatures values ranging from 16-20°C, have reported methane emission rates ranging from 3-6 g CH_4/capita·d. Thus, the reduction of gas emissions in the cold months was clearly observed. Conversely, as shown on Figure 2-11, in hot months an increase in gas release was observed. Kinnicutt also observed that gas production from septic tanks started almost immediately during warm months, while in cold months gas production was delayed until the tank contents warmed in the spring and summer. Thus, tanks that were emptied in the winter had the anaerobic digestion process inhibited by temperature.

Winneberger (1984) estimated 22- 28 L/capita·d of gas from a single septic system. Assuming that, as reported, the gas was 70% methane and a methane density of 0.67 g/L CH_4 results in an emission rate range from 14- 18 g CH_4/capita·d. Similarly, based on the method developed by the IPCC (1996, 2006), methane emissions from a domestic septic tank are approximately 25.5 g/capita·d (see Appendix C for calculations). However, as noted in Appendix C, the IPCC method uses an assumed methane conversion factor (MCF) of 0.5 to represent the conversion of organic matter to methane. A summary of the different methane emission rates estimates is presented in Table 2-9.

Table 2-9. Estimates of Methane Emission Rates from Septic Tank Liquid Surface.

Method	Year	Methane estimate ($g\ CH_4$/capita·d)
Kinnicutt et al.	1910	10.1[a]
Winneberger	1984	14 to 18[a]
COD loading	2009	11[b]
IPCC	2007	25.5[c]
Sasse	1998	18[d]

[a] Measured value.
[b] Calculated value assuming that 40 % of solids are removed as septage (see Appendix B).
[c] Calculated value assuming that half of the influent COD is converted anaerobically (see Appendix C).
[d] Calculated value assuming 25 % CH_4 dissolved (see Appendix D).

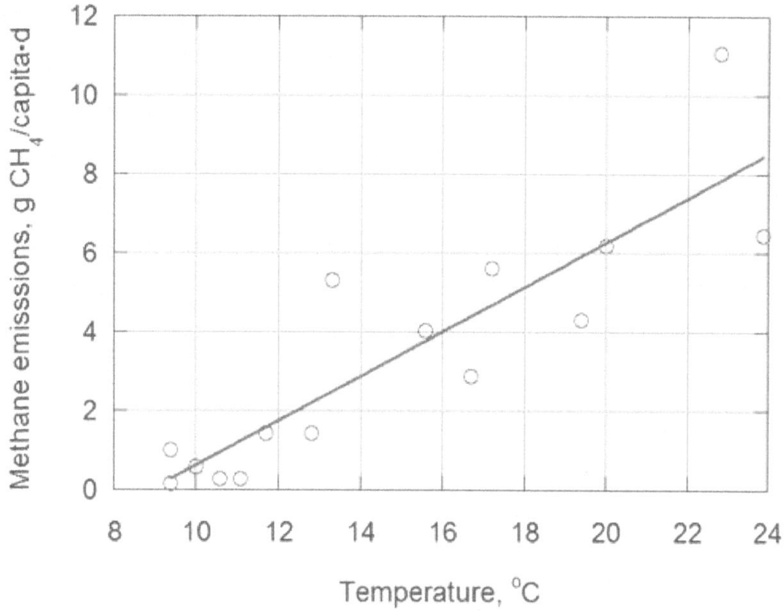

Figure 2-10. Relationship between Methane Emission Rates from Liquid Surface of Community Septic Tank and Water Temperature.
Kinnicutt et al., 1910.

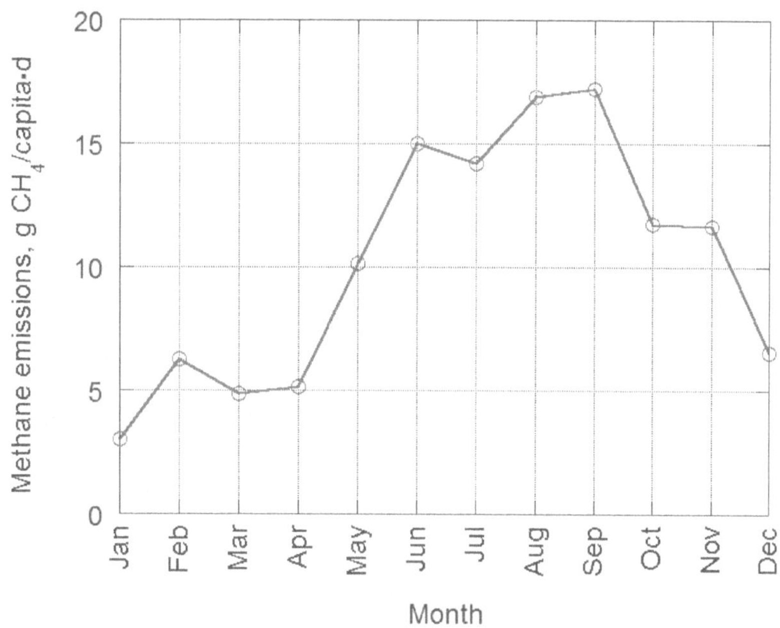

Figure 2-11. Monthly Methane Emission Rate from Liquid Surface of Community Septic Tank.
Kinnicutt et al., 1910

Sasse (1998) presented a model for estimating gas production from a septic tank system based on tank configuration, loading, temperature, and other variables. Using the Sasse (1998) model and typical values for North American septic tank design, loading, and configuration, a total methane production value of 18 g CH_4/capita·d is calculated (see Appendix D). The Sasse model also accounts for methane that leaves the tank in the effluent. For purposes of the calculation, it was assumed that 25% of the methane produced leaves the tank dissolved in the effluent.

2.4.5 Gases in Effluent Dispersal Systems

In soil-based effluent dispersal systems, wastewater is applied typically using a system of perforated pipes. At the point where septic tank effluent is applied to the soil, a clogging zone occurs as a result of high moisture content and excess organic matter. The clogging zone is composed of various materials, including wastewater particulate matter, microbial biomass, and inorganic precipitates. Based on oxygen diffusion rates alone, it has been determined that the supply of atmospheric oxygen is a limiting factor (Janna, 2007; Erickson and Tyler, 2001). Thus, the development of anaerobic conditions and clogging zones in conventional soil dispersal systems is an expected phenomenon.

2.5 Summary of GHG Emissions from Wastewater Systems

The increase in carbon dioxide and other gases in the atmosphere have motivated entities, such as the IPCC, to build GHG inventories to determine critical emitting sources. The published studies on GHG emissions from wastewater treatment plants and the relative importance of septic tanks gas emissions are summarized in this section.

2.5.1 Observations of GHG Emissions from Wastewater Treatment Plants

Concern about climate change has resulted in increased research on the emission of GHGs to the atmosphere (IPCC, 1996; U.S. EPA, 2006; Sahelli, 2006; Chandran, 2009; Foley and Lant, 2009). In general, onsite wastewater treatment systems have received less attention compared to full-scale wastewater treatment plants when accounting for GHG releases. However, it is important to have an understanding of the GHG estimated in these studies and the different approaches used to obtain them. Sahely (2006) used a life cycle assessment methodology to quantify GHG emissions from municipal wastewater treatment facilities in Canada, reporting carbon dioxide as the major gas contributing to GHG emissions, due to the predominance of aerobic treatment processes. It should be noted that life cycle assessment (LCA) studies are highly influenced by the boundary conditions and individual assumptions; consequently the findings should be considered only as a baseline for emissions inventories.

Foley and Lant (2009) published an experimental approach to evaluate gas fluxes from WWTPs in Australia. The study focused on the estimation of methane and nitrous oxide emissions from four full-scale treatment systems. The researchers pointed out that the estimated methane emissions from wastewater collection systems are underestimated and suggested that models should be developed to address this situation. Liquid methane measured at the inlet and outlet of various WWTPs are summarized in Table 2-10.

Cakir and Stenstrom (2005) developed a mass balance model to compare methane and carbon dioxide gases from aerobic and anaerobic wastewater treatment systems. The aerobic technology studied was a conventional activated sludge process and the anaerobic technology was an upflow anaerobic sludge blanket (UASB) reactor. It was reported in the study that aerobic processes release less GHG than anaerobic treatment processes for low strength (~300 g/L) influent BOD_u (ultimate carbonaceous oxygen demand or 20 d BOD). According to the model, for a BOD_u concentration of 100 mg/L, the dissolved methane in the effluent of the UASB was around 84 mg CH_4/L, while it was just 0.26 mg CH_4/L at the effluent of the aerobic process. The authors suggested that capturing the liquid methane produced after the UASB treatment was an alternative to reduce the total GHG emissions from the anaerobic treatment.

Table 2-10. Summary of Dissolved Methane Measured at Three WWTPs in Australia (Foley and Lant, 2009).

Wastewater system	Location	Flow m^3/d	Inlet, mg CH_4/d	Outlet, mg CH_4/d
Lagoons[a]	Adelaide Hills, South Australia	1,200	3.2 – 7.2[d]	0.3 ± 0.2
CAL[b]	Melbourne, Victoria	230,000	0.5 - 1.5[e]	0.7 ± 0.4
PST[c]	Sydney, New South Wales	275,000	0.5 – 1.5[e]	0.6 ± 0.1

[a] Uncovered anaerobic/facultative lagoons.
[b] Covered anaerobic lagoon.
[c] Primary sedimentation tank.
[d] Raw wastewater received via a rising main.
[e] Raw wastewater received by gravity.

2.5.2 Relative Importance of the Septic Tank Gas Emissions

Wastewater treatment systems are estimated to account for about 4% of the total methane emissions in the U.S. and are also a source of nitrous oxide, another GHG with an equivalent effect of more than 296 times that of carbon dioxide (IPCC, 2007; U.S. EPA, 2009). Using the IPCC methodology, the U.S. EPA (2009) has determined that a majority of the methane emissions associated with wastewater originate from onsite septic systems, due to the uncontrolled release of methane to the atmosphere and the large numbers of individual septic systems in use. According to the U.S. EPA (2010), methane and nitrous oxide emissions from all domestic wastewater systems are 15.7 and 4.9 Tg CO_2e/year, respectively. Using the U.S. EPA (2010) approach, the total methane emissions from septic systems is about 12.8 Tg CO_2e/year and nitrous oxide emissions are not accounted for. Methane and nitrous oxide emissions attributed to centralized wastewater treatment are about 2.9 and 4.9 Tg CO_2e/year, respectively. It should be noted that energy and chemicals used in collection and treatment, as well as gas emissions associated with wastewater collection, are not accounted for in the U.S. EPA (2010) emissions model.

Given the data presented in Table 2-9, there is a significant difference in the previously measured methane emission values compared to the IPCC model estimates; thus, there is reasonable justification to conduct additional studies to quantify the amount of gases released to the atmosphere from septic tanks. The U.S. EPA (2010) model used to determine GHG emissions from septic systems applies the same methane correction factor, MCF (see Appendix C) used in the IPCC model, therefore both models result in the same estimate of methane emissions (see Table 2-9). Again, the MCF value of 0.5 used in the U.S. EPA (2010) model is based on the assumption that half of the influent organic matter is converted to methane.

CHAPTER 3.0

Materials and Methods

The topics presented in this section include the design of the devices and protocols used for sampling at the (a) septic tank liquid surface, (b) venting system, and (c) soil dispersal system; GHG laboratory analysis methods and procedures; factors taken into account for data quality control and data acquisition; and equations used for data analysis.

3.1 Design of Sampling Devices

This section presents a description of the flux chamber designed for use in the septic tanks, the modifications made to the flux chamber for use in the soil dispersal systems, and the device designed to obtain gas samples from the venting system.

3.1.1 Flux Chamber Design for Use in Septic Tanks

Measurements of GHG emissions from soil-plant systems using flux chambers and the corresponding sampling techniques are well established (Hutchinson and Livingston, 1993). However, little information is available on gas flux measurements from the liquid/solid surface of septic tanks. Thus for this research, a flux chamber design based on the soil-plant system was modified, constructed, and tested, along with the development of a corresponding sampling methodology. The flux chamber is shown on Figure 3-1.

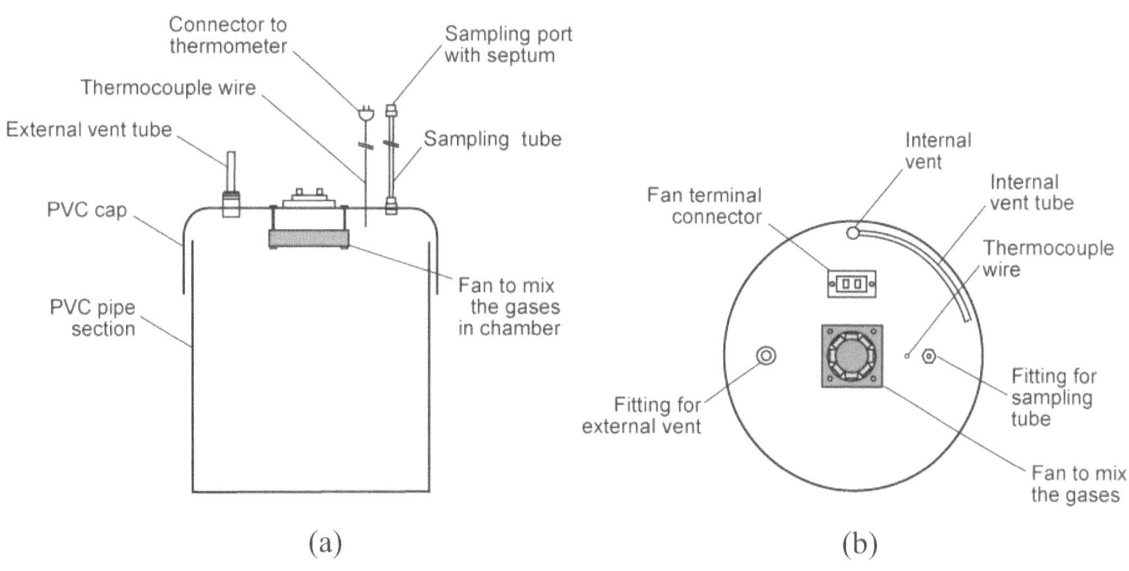

Figure 3-1. Flux Chamber Developed to Sample Gases from Liquid Surfaces: (a) Side View of the Flux Chamber and (b) Internal View of the Flux Chamber Where the Fan and Internal Vent are Visible

The main body of the flux chamber was constructed from a PVC pipe section (12" diameter, 12" length), inserted into a 12" diameter PVC cap. A 12-volt fan was installed inside of the cap to ensure that the gases were well mixed in the chamber. A 3/16" brass Swagelok fitting was inserted into the PVC cap to hold 4' of Teflon tubing (3/16" external diameter). A second brass piece at the end of the tubing was fitted with a septum to comprise the sampling port used to withdraw the gas samples. Two additional vinyl tubing vents were installed, one being a 1/4" internal vent that extended 8" inside of the cap and one a 1/2" diameter external vent of 6' in length with a valve at the end as seen on Figure 3-1b.

The purpose of the vents was to maintain atmospheric pressure inside the chamber. The larger vent was only used while initially submerging the flux chamber into the liquid to account for the large amount of air that needed to be displaced. The larger vent was closed after the flux chamber was in position for sampling. The total gas volume inside of the flux chamber during sampling was determined using a scale fixed to the side of the chamber; the scale was used to measure the depth of the chamber submergence. The chamber was suspended in the tank from an above ground tripod using chains attached to the cap.

3.1.2 Flux Chamber Inserts for Septic Tanks

A tank insert designed and used to prevent disturbance of the septic tank contents and to support the flux chamber when measurements were being taken is shown on Figures 3-2 and 3-3. This insert was necessary when there was a scum layer present on the liquid surface. The insert was composed of a 10" diameter PVC pipe section with a length of 8" with a channel at the top. The insert was supported with PVC legs going to the bottom of the tank, such that the insert was submerged about 6" into the liquid in the tank. The channel was filled with water prior to sampling. The flux chamber was lowered into the water filled channel to seal the contents of the chamber. The tank inserts were left in place for the duration of the experiment. Some tanks had limited access and it was not possible to fit the 12" flux chambers inside the risers; therefore a 6" diameter flux chamber and inserts of the same diameter were built to address this situation. A summary of the chamber size use per site and compartment is shown in Table 3-1.

Table 3-1.Chamber Size Used in Each Site by Septic Tank Compartment.

Septic tank compartment	Diameter of flux chamber used at each site (inch)							
	1	2	3	4	5	6	7	8
First	6	6	6	6	12	6	6	6
Second	6	6	12	12	12	6	6	6

(a) (b)

Figure 3-2. Example of Insert Used for Gas Sampling from Septic Tanks with Scum Layer: (a) View of 6 and 12 inch Inserts and (b) Close-up View of the 12 inch Insert.

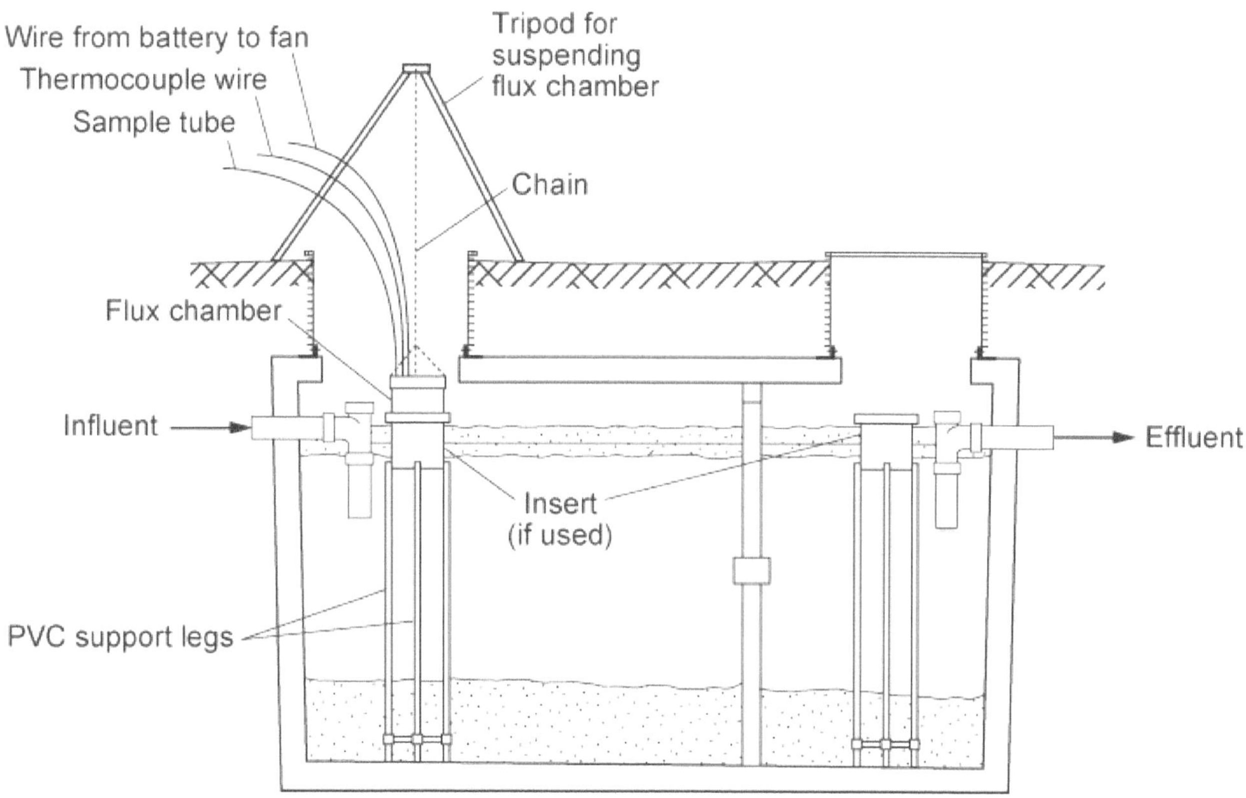

Figure 3-3. Sampling Device Used to Obtain Gas Samples Through Scum Layer. Note the Placement of the Flux Chambers and Inserts.

3.1.3 Flux Chamber Design for Use in Soil

A chamber similar to that described in Section 3.1.1 was constructed to take samples from the leach field (Hutchinson and Livingston, 1993, USDA-ARS, 2003). The key difference between the leach field and septic tank flux chambers is that the leach field chamber did not include the external vent and it had an additional covering of reflective insulation (as seen on Figure. 3-4) to decrease the sensitivity of the measurements to radiant heating.

Figure 3-4. Flux Chamber Designed for Testing Gases from the Leach Field.

3.1.4 Flux Chamber Inserts for the Soil

The inserts for the soil dispersal system were made of PVC pipe (12" diameter and 4" length). The soil inserts were functionally similar to the scum layer inserts described above. Six inserts were installed above each soil dispersal system a week in advance to the first sampling event in the soil. The insert pieces were left in the soil over the entire sampling period to minimize disturbances of the soil when the flux chambers were set and the samples taken (USDA-ARS, 2003).

3.1.5 Vent Sampling Device

An apparatus was built to sample vented gases from the septic system cleanout port. As shown in Figure 3-5, the sampling device consisted of a 3 or 4" PVC slip cap and threaded ABS adapter as needed to fit the cleanout port. Teflon tubing and a hot wire anemometer with a telescoping handle were mounted on the cap. A 3/16" brass Swagelok fitting was used to hold the Teflon tubing in place. A second brass piece at the end of the tubing was fitted with a septum for extraction of the gas samples with a syringe. A 1/2" cord-grip connector was attached to the cap to hold the hot wire anemometer (See Figure 3-5). Lengths of 1/2" PVC pipe sleeves were attached to the bottom of the cap with threaded adapters and used as guides for the sampling tube and the hotwire anemometer. The length of the PVC pipe sleeves varied depending on the depth to the same location at the centerline of the drain line as seen in Figure. 3-5. The device fit tightly in the clean out port, creating a tight seal. Because the cleanout was completely sealed when the gas sampling apparatus was in place, gases were constantly moving through the drain line of the house and exiting the building vent as would occur under normal conditions. The sample tubing and anemometer were small relative to the cross-sectional area of the drain line and therefore not expected to impact the gas flow.

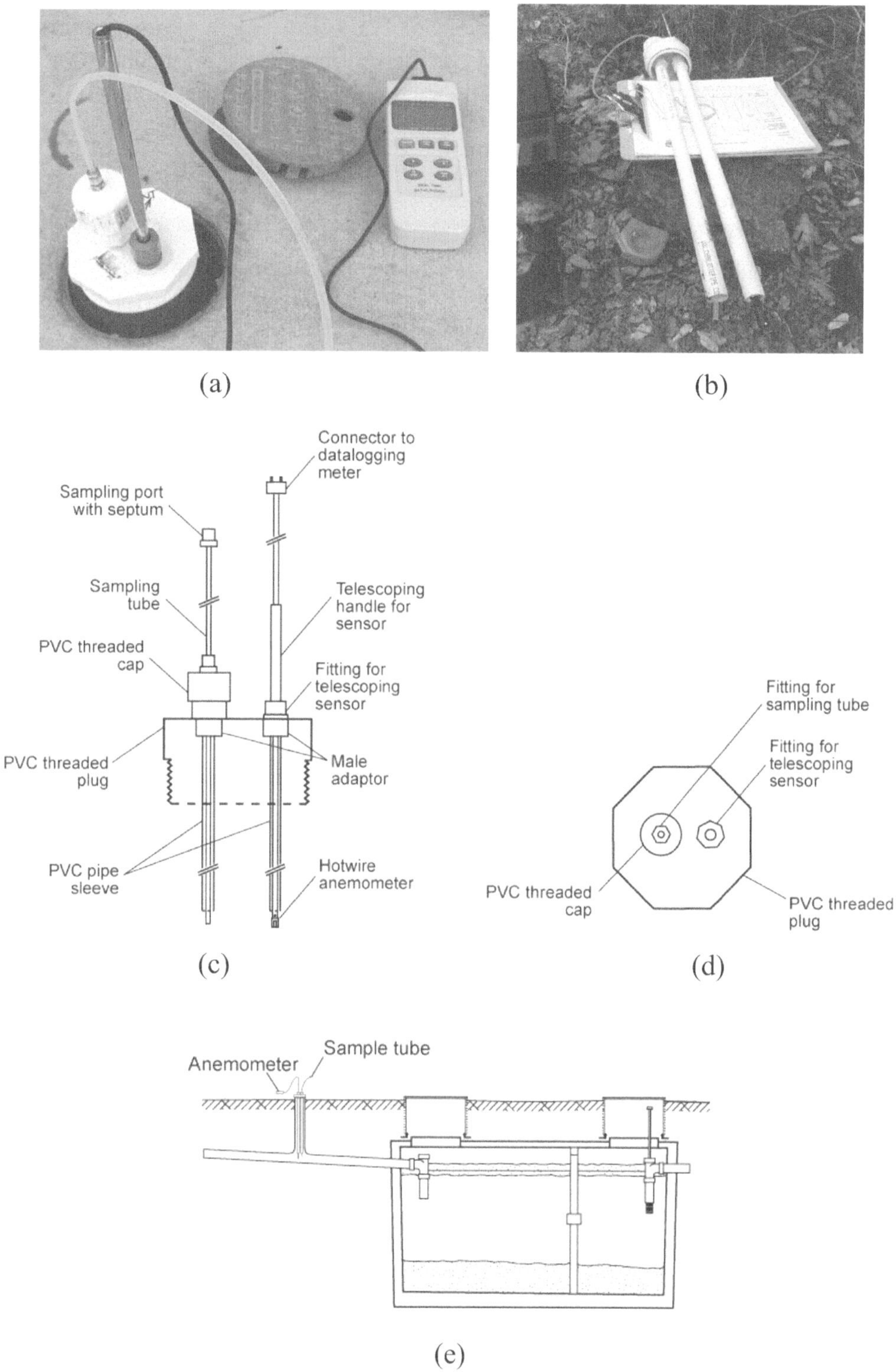

Figure 3-5. Device Developed to Sample Gases from the Cleanout Vent (a) Main Body of the Device and Anemometer (b) View of the PVC Pipes Sleeves, (c) Profile and (d) Plant of the Vent Sampling Device. Not at Scale, and (e) Illustration of Technique Developed to Sample Vent Gases.

3.2 Sampling Protocols

The three principal components of an onsite wastewater treatment system were individually sampled to determine the GHG emissions. Each component had specific characteristics requiring a different sampling method. The following section contains descriptions of the sampling method developed for the septic tank liquid surface, the venting system, and the soil dispersal system. In addition, the technique used to obtain samples of dissolved methane is described.

3.2.1 Sampling Method for Liquid Surfaces

To sample at the liquid surface, the septic tank access port was opened and a tripod was set up to suspend the flux chamber (see Figure 3-6). The sample was obtained by submerging the chamber in the liquid. If a scum layer was present, an insert was installed prior to obtaining samples (see Section 3.1.2, Figure 3-2 and Figure 3-3).

1. The flux chamber was slowly submerged partially into the water. The chamber depth was recorded in every sampling event to allow later calculation of chamber volume. A complete water seal around the chamber was maintained to ensure accurate flux measurements. When the chamber was put into the water, the timer was started.

(a) (b)

Figure 3-6. Use of Flux Chamber for Gas Sampling from Septic Tank Located in Cool, CA (a) Deployment of Flux Chamber Into First Compartment of 1200 Gallon Septic Tank and (b) Extraction of Headspace Sample from Flux Chamber Using Syringe.

2. A 12 mL syringe was inserted into the sampling tube septum and 6 mL of volume was taken and discharged into the air (see Figure 3-6). This step was used to purge the accumulated gases in the sample line, which had a volume of 6 mL. Simultaneously, a temperature reading was obtained.
3. After purging the sample line, two 12-mL samples were collected and put into one of the previously evacuated vials to obtain a total sample volume of 24 mL (vial evacuation

procedure is shown in Appendix E). The vial was labeled and protected from the sun. For the initial (t = 0) sample, the fan was not needed because a concentration gradient had not developed.

4. The time interval between samples was 10 min. Before withdrawing further samples, the fan was energized briefly (5 sec) to mix the contents of the flux chamber. At time 10 min, the fan was turned off and the second sample was collected after purging the sample line again. Again the temperature inside the chamber was recorded.
5. Step 4 was repeated for the third sample. The flux chamber was then taken out of the water and the access port closed and secured.

3.2.2 Sampling Method for Soil Systems

1. Five 12 inch PVC permanent inserts were inserted at different locations in the leach field area and another one was placed outside the leach field to use as a control site. The permanent inserts were used to hold the flux chambers while gas measurements were taken; the inserts remained in place for the duration of the study. These pieces have a similar design to the insert used for sampling through the scum layer (Figure 3-7); however they were shorter (6 or 4" in length). As with the scum layer sampling, a water seal was applied, creating a complete seal between the insert and the flux chamber.

(a) (b)

Figure 3-7. Use of Flux Chamber in the Soil Dispersal System (a) Permanent Insert and (b) Extraction of the Headspace Sample from the Flux Chamber.

2. The flux chamber was placed on the inserts sealing them together and avoiding any gas leakage during sampling.
3. The 12 mL syringe was inserted into the sampling septum and used twice to withdraw a total sample volume of 24 mL at time zero; the overall process was similar to the liquid surface measurements described above. After discharging the samples into a previously evacuated vial, a temperature reading was obtained.
4. The time interval between samples was 10 min. Before withdrawing the next samples, the fan was energized briefly (5 sec) to mix the contents of the flux chamber. At time 10 min, the fan was turned off and the second sample was collected. Again the temperature inside the chamber was recorded.
5. Step 4 was repeated for the third sample after which the flux chamber was taken out of the insert piece.

3.2.3 Sampling Method for Vent System

1. The cap of the cleanout port located before the septic tank was removed to verify that there was no water flowing through the pipe to avoid damaging the hot wire anemometer.
2. The length of the anemometer as well as the sampling line was adjusted to make sure that it was placed exactly in the middle of the pipe (see Figure 3-5).
3. The sampling device was installed in the cleanout port and the anemometer was turned on.
4. The 12 mL syringe was inserted into the sampling septum (Figure 3-8) and used twice to withdraw a total sample volume of 24 mL into an evacuated vial, as above. An air velocity reading was made each time a sample was taken.
5. Two more samples were taken following step four with a sampling interval of two min between them. Finally, the sampling device was removed and the cleanout cap was replaced.

Figure 3-8. Sampling of Gases from the Venting System.

3.2.4 Sampling Method for Aqueous Methane

1. Plastic tubing attached to a 12 mL syringe was inserted to approximately the middle of the liquid column in the first compartment of the tank.
2. Two full syringes of wastewater were withdrawn to purge the tube line. A final 12 mL sample was taken. Five mL of this sample were inserted into an evacuated vial. Then the syringe was pulled out leaving the hypodermic needle inside the septum for 30 sec to equilibrate the sample to ambient pressure (Alberto et al., 2000). A second 5 mL sample was drawn.
3. The plastic tubing was then inserted in the middle of the liquid column of the second compartment and step two was repeated.
4. After returning the samples to the laboratory, all the samples were shaken for 24 hours (Guisasola et al., 2008) to let the methane equilibrate between the liquid and gas phases.

5. The headspace gases were extracted from the vial using a syringe with a needle and placed in an evacuated tube for later GC analysis.

3.2.5 Sampling Method for Water Temperature

Thermocouples were attached at 12" intervals to a PVC pipe (1" external diameter with a total length of 7'). After the gas samples from the liquid surface were taken the following steps were performed.

1. The pipe was carefully inserted in the first compartment of the septic tank.
2. The thermocouple connectors were plugged to a thermometer reader and the temperature readings were recorded.
3. Steps 1 and 2 were repeated for the second compartment.
4. The pipe was pulled out from the second compartment and rinsed with water.

3.2.6 Sampling Method for Water Quality

A Myron L Ultrameter II™ was used to measure dissolved solids, pH and redox potential. The water quality measurement procedure for these three parameters was as follows:

1. Vinyl tubing attached to a 20 mL syringe was inserted into the middle of the water column in the first compartment of the septic tank.
2. The syringe was purged with liquid from the tank twice. The liquid was returned to the septic tank.
3. The syringe was used to place liquid into the Ultrameter II™ cup cell after which the reading was recorded. The liquid was poured back into the septic tank.
4. The syringe was filled again and 10 mL of wastewater was poured into a 12 mL plastic vial taken back to the laboratory for COD analysis. A HACH DR-890 colorimeter was used for the COD analysis, applying a COD digestion method. (U.S. EPA, 1993)
5. Steps 1 to 4 were repeated for the second compartment of the septic tank.

3.2.7 Sludge and Scum Thickness

1. A Sludge Judge sampling device (Nasco Equipment, Inc., Fort Atkinson, WI) was inserted until it reached the bottom of the septic tank.
2. The sampling device was pulled back from the tank bottom and the sludge thickness was recorded (see Figure 3-9).
3. The scum thickness was estimated using the same sampling device and feeling the scum thickness using the end of the probe. The measurement was recorded.
4. The sampling device was rinsed with water.

3.2.8 General Observations

Field observations were recorded during every sampling event on previously prepared forms (see Appendix F) designed for each system component. General observations on the septic tank scum appearance, presence of invertebrates, changes in the wastewater color and distinctive episodes such as turnover events and laundry water discharges were the most common aspects observed.

3.3 Gas Analysis

The gas samples were analyzed by a Shimadzu gas chromatograph (Model GC- 2014) with a 63Ni electron capture detector (ECD) linked to a Shimadzu auto sampler (Model AOC-5000). The samples were analyzed for CH_4, CO_2 and N_2O. The autosampler uses a gas-tight syringe to remove 5 mL gas from a sample vial and inject it into the GC port. The instrument was operated by experienced technicians with specialized training using a well established analytical protocol at the geochemistry laboratory of the UC Davis Plant and Environmental Science Department.

Figure 3-9. Sludge Measurement Taken at Site 4. The Red Marks are Located Every Foot.

3.4 Quality Control for the Samples

Before each sample event, one person from the research team was in charge of evacuating the vials and placing a mark on the vial's cap each time the vial was evacuated. After three subsequent evacuations, the septum inside the cap was replaced. All vials were labeled with the date and a code for each site prior to the sampling event (i.e., Septic Tank 1, inlet, time = 0 was labeled ST1 in 0).

During each sampling event one person from the research team was designated to be in charge of handling the samples taken on that particular day including protecting them from sun exposure, breakage or damage and delivering them to the laboratory for analysis.

The laboratory technicians were provided with three days of advance notice of the sampling event via e-mail to ensure that the GC was available and working properly. The laboratory technicians were responsible for measuring the CO_2, CH_4, and N_2O concentrations in the samples as well as the gas standards (two standards of each concentration per 24 samples) and submitting the results to the project manager. The system was calibrated daily using analytical grade standards (Airgas Inc., Sacramento CA). The quality of the samples was insured by using controls treated (age and storage conditions) the same as the field samples. Sample collection in the field and analysis of samples by GC was performed according to clearly established protocols.

3.5 Data Analysis

To calculate the gas fluxes from each component of the septic system, the data generated in the GC was analyzed based on different approaches. The calculation approach used in the septic tank, soil dispersal system and venting system is presented in this section.

3.5.1 Data from Septic Tank Analysis

The septic tank data were analyzed based on an algebraic linear model describing the gas flux rate coming into the headspace of the flux chamber. The gas flux was calculated using the slope of the trace gas accumulation curve from the measurements taken at each compartment of the septic tank during the sampling event. A sample plot for methane and carbon dioxide showing the linear fit for one flux measurement is presented in Appendix G. The equations used to calculate the septic tank flux values are shown below in Eq- 3-1 and Eq. 3-2. As shown in Eq. 3-1 the concentration in ppm (raw data from laboratory) were first converted to concentration in mg/m^3.

$$Gas\ concentration\,(mg/m^3) = \frac{(C_{ppm}/10^6)(MW)(1000\ mg/g)}{\left(\frac{RT}{P}\right)} \qquad \text{(Eq. 3-1)}$$

Where C_{ppm} is equal to concentration in ppm, MW is the molecular weight of the gas under consideration (g/mol), R is the gas constant (0.000082057 $atm \cdot m^3/mol \cdot K$), T is the absolute temperature (K), and P is the absolute pressure of the gas (atm). The concentration values (in mg/m^3) from individual measurement events (typically 4 to 5 consecutive samples) were then plotted as a function of time (See Appendix G). The slope m, in units of $mg/m^3 \cdot sec$ derived from a linear fit of the data is then used to compute the flux using Eq. 3-2.

$$Flux\,(g/capita \cdot d) = \frac{(m)(V_{FC})(A_{comp})(86400\ sec/d)}{(1000\ mg/g)(SA_{FC})(capita)} \qquad \text{(Eq. 3-2)}$$

Where m is the slope of a linear fit to the gas concentration data ($mg/m^3 \cdot sec$), V_{CF} is the volume of the flux chamber (m^3), A_{comp} is the liquid surface area of the compartment of the septic tank where the sample was taken (m^2), SA_{FC} is the liquid surface area occupied by the flux chamber (m^2), and capita is the number of occupants in the house.

3.5.2 Data from the Soil Dispersal System

A non-linear model (Hutchison and Livingston, 1993) was applied to account for gas production and consumption in the soil based on diffusion theory. The concentration values in ppm (raw data from laboratory) were first converted to concentration in mg/m^3 using Eq. 3-1. Equation 3-3 was then used to determine the gas flux rate from the soil dispersal system.

$$Flux\,(mg/m^2 \cdot d) = \frac{V(C_1 - C_0)^2}{A \cdot t(2C_1 - C_2 - C_0)} \ln\left(\frac{C_1 - C_0}{C_2 - C_1}\right) \quad for\ t_2 = 2t_1\ and\ \frac{C_1 - C_0}{C_2 - C_1} > 1 \qquad \text{(Eq. 3-3)}$$

Where V is the volume of the chamber (m^3), A is the soil surface area occupied by the flux chamber (m^2), C_o, C_1 and C_2 are the gas concentrations (mg/m^3) at times t_0, t_1, and t_2, and t is the time interval between t_0 and t_1, and t_1 and t_2 (d). In this case, the concentrations were measured three times during the same interval length t (i.e., every 10 min).

3.5.3 Data from the Vent System

The concentration values in ppm (raw data from laboratory) were converted to concentration in mg/m^3 using Eq. 3-1. The air velocity inside the pipe was measured using a hot wire anemometer. The air flow rate was calculated using Eq. 3-4.

$$Flow\ rate\ (m^3/d) = \pi r^2 \cdot v \qquad \text{(Eq. 3-4)}$$

Where, r is the radius of the pipe (m) and v is the velocity inside the pipe (m/d). The mass flow of gas constituents was estimated as the product of the measured gas concentration (mg/m^3) and the measured flow rate.

CHAPTER 4.0

DESCRIPTION OF FIELD SITES

The selection criteria for the sites and septic tanks, the general characteristics of the sites chosen, a detailed description of each site septic system, the experimental approach for sampling at the septic tank, venting, and soil dispersal system and the preparation for field sampling are presented in this section.

4.1 Site and Septic Tank Selection

The Georgetown Divide Public Utility District (GDPUD) is the entity in charge of the wastewater management in the Auburn Lake Trails (ALT) development. The development is located in El Dorado County, next to highways 49 and 193, one mile from the town of Cool. ALT has development rights for the construction of 1,100 parcels; at present 999 lots have been developed (GDPUD, 2010). The GDPUD has a detailed inventory of the septic tanks located in Auburn Lake Trails development (38° 54' 51.48"N, 120° 57' 08.85"W) as well as records related to septic tank capacity, date of installation, pump out intervals, and maintenance. Based on the GDPUD information, a preliminary inspection of the septic tanks was conducted; seven tanks met the selection criteria for this study. An eighth tank, located in Davis, California, was also included in the study to perform 24-hour sampling events to capture the variability of GHG emissions throughout the day. The criteria used to select the eight septic tanks are described below:

- Site was readily accessible
- Tank access lids were easy to open/close
- Tanks were structurally sound, water-tight, and gas-tight
- The flow inside the vent system pipe was measurable
- The soil dispersal systems had appropriate land inclination for setting up sampling equipment

Following a series of preliminary gas emission measurements from all sites included in the study, several sites were selected for further evaluation based on accessibility and performance characteristics similar to a typical system, as discussed below.

4.2 General Site Characteristics

Sites 1 to 7 were located in the ALT development in Cool, CA, and Site 8 was located in Davis, CA. Gas samples from the septic tank liquid surface were taken at all the sites. Sites 1, 2, and 7 were selected for a more detailed study at the venting and soil dispersal systems and at Site 8, two 24 hour sampling events were performed.

All the systems were less than 15 years old, except for the septic tank at Site 8, which was built in the 1940s. All the septic systems were gravity flow systems, comprised of a septic tank and soil dispersal system, with no other advanced treatment. It should be noted that the onsite system at Site 8 also included a graywater system for the laundry water. All the septic tanks were double compartment concrete tanks and their capacity ranged between 1000 and 1250 gal. Sites 5, 6, and 7 were the only sites with effluent filters. It was noticed that the septic tank lids at Sites 1 and 7 did not seal as well as the other septic tanks. The first compartments of the septic tanks at Sites 2, 3, and 7 had well developed scum layers that ranged from 1 to 5". The first compartment of the septic tanks at Sites 1, 4, 5, 6, and 8 had only patchy or thin scum layers. A scum layer was never present in the second compartment of the septic tanks.

4.3 Description of the Individual Site Characteristics

A survey was given to each household, which included general questions such as number of occupants, water saving fixtures in the house, number of bathrooms, use of garbage grinder, and questions related to the septic systems such as pumping intervals and age of the system. More detailed information such as monthly water consumption, inspections, and maintenance records were provided by the GDPUD. The results of the survey and the information provided by the GDPUD are presented in Table 4-1.

Table 4-1. General Characteristics of the Septic Tanks in the Study Group.

	Septic tank number							
Characteristic	1	2	3	4	5	6	7	8
Location	ALT Development, Cool, CA							Davis, CA
Number of occupants	2	2	2	2	2	2	3	4
Water saving fixtures	LFS, LFT[a]	LFS, LFT	None	None	LFS, LFT	LFT	LFS, LFT	Greywater[b]
Number of bathrooms	2.5	3	3	2	2	2	4	4
Use of garbage grinder	Occasional	Rarely	Rarely	Rarely	Almost never	Almost never	Once/wk	No
Years since septic tank pumped	3	Never	Never	0.5	12	Never	3	>12
Date built	unknown	2005	1989	unknown	unknown	2002	unknown	1940s
Water consumption (L/capita·d)[c]	243	604	461	1345	411	84	170	180

[a] LFS = low flow shower, LFT = low flow toilet.

[b] Laundry water diverted to greywater system.

[c] Average water usage based on winter season (November through February) 2008 - 2010.

A plan view of the septic system including the septic tank, cleanout vent, and soil dispersal inserts along with a detailed description of the physical characteristics of the septic tanks such as the tank volume, capacity, and the inserts installed at each compartment to support the gas flux chambers during the sampling events are presented in Table 4-2.

Table 4-2. Physical Characteristics of the Eight Septic Tank Systems Used in the GHG Emissions Study. All Site Plans are not to Scale and Oriented with North Towards Top of Page.

<table>
<tr><th>Site</th><th>Site plan</th><th>System description</th></tr>
<tr><td>1</td><td>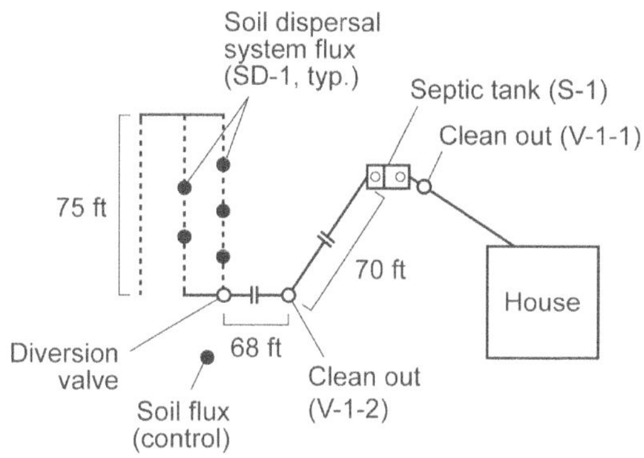
</td><td>The septic tank, located in a shady area, has a capacity of 1200 gal. An insert was deployed in the first compartment of the septic tank to support the 6" flux chamber used to obtain the gas samples. It was not necessary to use an insert in the second compartment; the flux chamber was always deployed directly on the liquid surface.</td></tr>
<tr><td>2</td><td>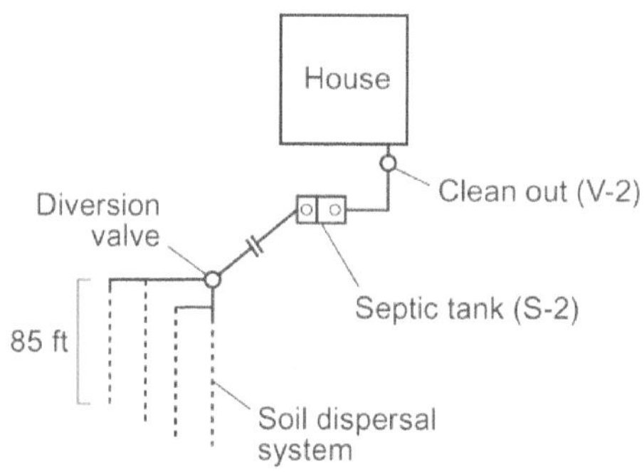
</td><td>The septic tank, located in a sunny area, has a capacity of 1200 gal. An insert was installed in the first compartment of the septic tank to hold a 6" flux chamber. It was not necessary to install an insert in the second compartment; the flux chamber was always deployed directly on the liquid surface.</td></tr>
<tr><td>3</td><td>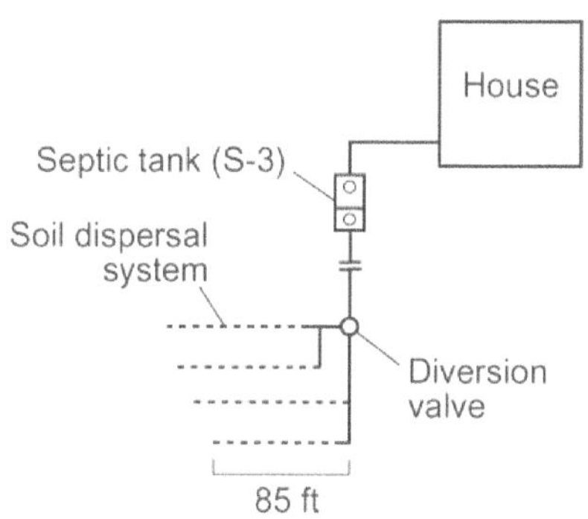
</td><td>The septic tank, located in the shade, has a capacity of 1250 gal. In the first compartment an insert to support a 6" flux chamber was installed. It should be noted that in the second compartment a 12" flux chamber was used to take the gas samples and it was deployed directly on the liquid surface.</td></tr>
<tr><td>4</td><td>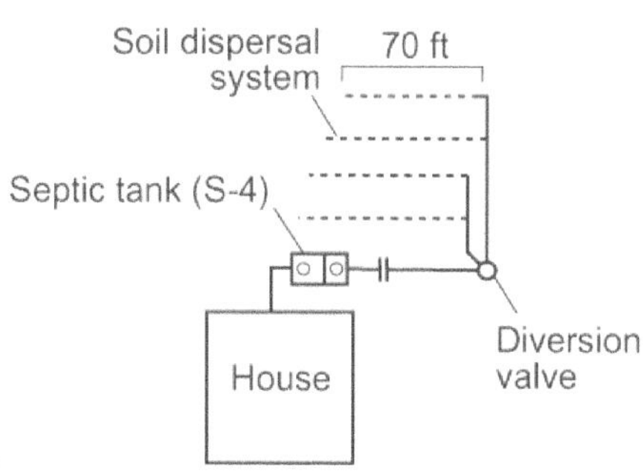
</td><td>The septic tank, located in a sunny area, has a capacity of 1000 gal. Inserts were not installed in this tank because it did not have a scum layer formation and therefore the 6 and 12" flux chambers used for the first and second compartment, respectively were always deployed directly on the liquid surface.</td></tr>
</table>

Continued on following page

Table 4-2. Continued from previous page.

Site	Site plan	System description
5	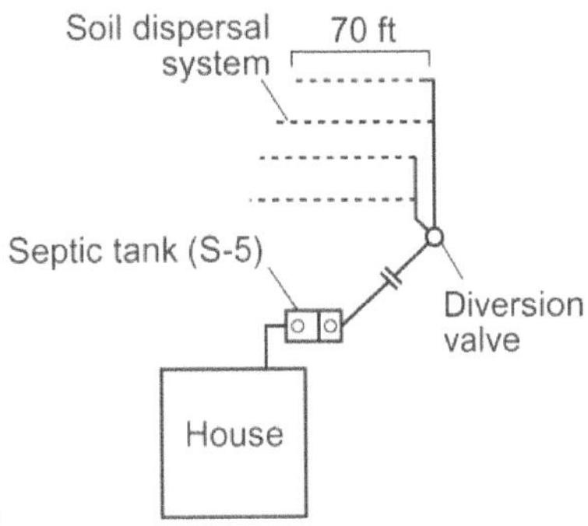	The septic tank, located in a sunny area, has a capacity of 1250 gal. Inserts to hold 12" flux chambers were installed in both compartments of the septic tank.
6	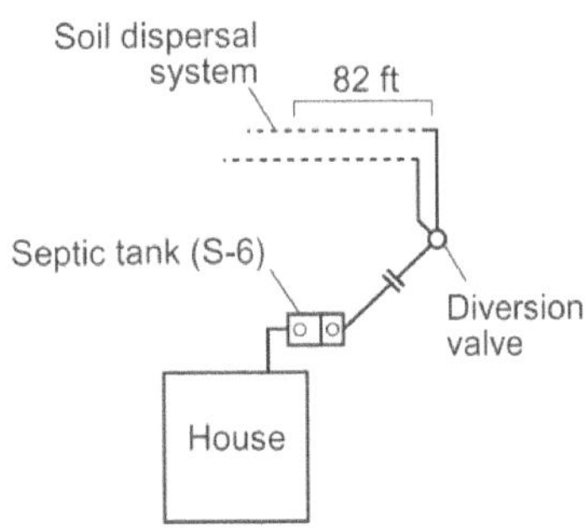	The septic tank, located in a partially shaded area, has a capacity of 1200 gal. Inserts to support 6" flux chambers were installed in both compartments.
7	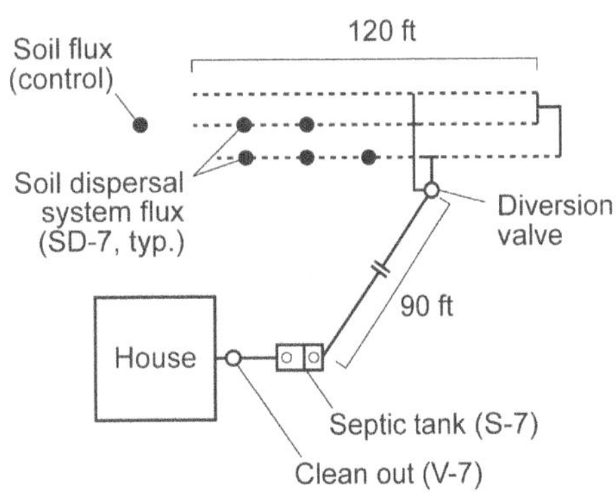	The septic tank, located partially in the sun, has a capacity of 1200 gal. Inserts to support 6" flux chambers were installed in both compartments.
8	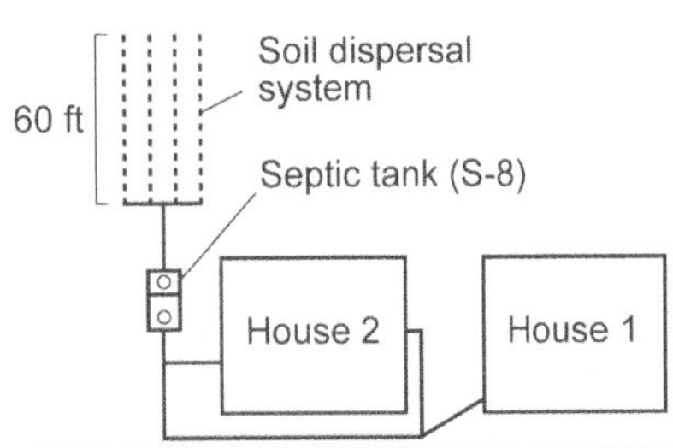	The septic tank, located in a partially shaded area, has a capacity of 1200 gal. Inserts to support 6" flux chambers were installed in both compartments of the septic tank. Site 8 was the only septic tank located in Davis, CA and it was selected to perform two 24-hr sampling events. A sanitary tee was not installed in the inlet.

4.4 Sampling Schedule

Gas flux measurements from all eight tanks were taken at various times over a four month period (September to December 2009) to attempt to capture the temperature effect on the GHG emissions. Due to external time constraints placed on this project, sampling could not be continued after December. The flux measurements and gas samples were taken in the septic tank, venting system and soil dispersal system. Each event included measurement of the gas flux and concentration of CH_4, CO_2 and N_2O. In addition, several water quality parameters (pH, dissolved solids, redox potential, COD, and aqueous methane concentration) were also measured in the inlet and outlet chambers of the septic tank. A summary of the sampling events, detailing the date, frequency, location and the parameters measured is provided in Table 4-3.

Table 4-3. Sampling Type, Location, and Frequency.

	Site[a,b]							
Date	1	2	3	4	5	6	7	8[c]
9/24	S'U,T	S,U,T	S,U,T	S,U,T	S,U,T	S,U,T	-	-
10/07	S,T	S,T	S,T	S,T	S,T	-	S,U,T	-
10/21	-	-	-	-	-	-	-	S,U,T
11/05	S,T,W	S,T,W	S,T,W	S,T,W	S,T,W	S,T,W	S,T,W	-
11/10	S,V,T	-	-	-	-	-	-	-
11/12	S,V,D,T, M	-	-	-	-	-	-	-
11/17	-	-	-	-	-	-	S,V,D,T, M	-
11/19	S,V,T,W, C,M	S,V,T,M	-	-	-		-	-
12/01	-	-	-	-	-	-	S,V,D,T, W,C,M	-
12/03	S,V,D,T, W,C,M	-	-	-	-	-	-	-
12/08	-	S,V,T,W, C,M	-	-	-	-	-	-
12/10	-	-	-	-	-	-	-	S,T,W, C,M
1/5	V, W							

[a] At each sampling event, three to six gas samples were taken from the septic tank (at each compartment), venting system (before and after the septic tank when possible), and soil dispersal system, identified as S,V and D, respectively. Each sample included measurement of the gas flux and concentration of CH_4, CO_2 and N_2O.
Sludge and scum thickness measurements are identified as U.
Water temperature measurements are identified as T.
Water quality measurements of pH, dissolved solids and redox potential are identified as W.
Dissolved methane measurements are identified as M.
COD measurements are identified as C.
[b] Sites 1 to 7 located in Cool, CA.
[c] Site located in Davis, CA.

Three of the septic tanks that appeared to have similar average gas emission rates (i.e., Sites 1, 2, and 7), and had accessible venting and soil dispersal systems were selected for additional gas flux measurements from the venting and soil dispersal system (see Table 4-2). This additional sampling was done with the intent of developing a mass balance on the overall methane emission rates from the system. The mass balance analysis, presented in Chap. 5, is based on data from Sites 1 and 7, where flux values from the septic tank, soil dispersal system, and vent system were obtained. The vent samples from Site 2 were used in developing the overall atmospheric emissions; however, because soil dispersal system flux was not determined at this site, it was not used in the mass balance analysis.

Flux chambers were designed and built to obtain flux measurements at the liquid surface and above the soil dispersal system. A special device was constructed to obtain flux measurements from the venting systems. The experimental apparatus and methods developed to sample from the three components (liquid surface, gas vent, and soil dispersal system) are described in Sections 3.1 and 3.2, respectively.

4.5 Preparation for Field Sampling

Preparations for each sampling day were made at least one day in advance. Before sampling, the use of personal protection equipment (PPE) was verified for all participants to avoid direct contact with the wastewater. PPE consisted of glasses, gloves and closed toe shoes. To assure a smooth and accurate sampling process, two to three people were required at each sampling event. Parameters measured in the field, comments and a full description of each site were recorded on sampling forms (see Appendix F) designed according to the component of the septic system (i.e., septic tank, soil dispersal system or vent system) analyzed.

Sampling vials (24 mL) were evacuated no later than one day prior to sampling. Two control vials with methane concentrations of 10 and 100 ppm, two controls with carbon dioxide concentrations of 1,000 and 10,000 ppm and two controls with nitrous oxide concentrations of 1.12 and 5.02 ppm were prepared in vacuumed vials to be taken to the field and analyzed along with the samples collected in the field.

CHAPTER 5.0

RESULTS AND DISCUSSION

The results from the field studies of the gas emissions from septic tanks are presented and discussed in this section. The specific topics include: 1) the GHG emission rates from septic systems, 2) mass balance analysis and 3) the sources of variability in gas emission rates. A summary of all data collected in the study is presented in Appendix H.

5.1 GHG Emission Rates from Septic Systems

Gas emissions from septic systems may be composed of gases produced in the septic tank, soil dispersal system, and drain piping. The items discussed below include: (a) specific gas emission rates from the septic tank measured with flux chambers, (b) composite gas emission rates estimated from sampling of the venting system, (c) gases present in septic tank liquid samples, (d) gas emission rates from the soil dispersal systems measured using flux chambers, and (e) comparison of methane emissions models

5.1.1 GHG Emission Rates from Septic Tanks as Measured Using Flux Chambers

The GHG emissions that were measured in this study include methane, carbon dioxide, and nitrous oxide. Eight septic tanks were sampled using 6 and 12" flux chambers. The gas bubble pattern in the septic tank is assumed to be random. Therefore, with a sufficient number of samples the statistical distribution of gas flux values should be identical, independent of whether the 6 or 12" flux chamber size was used.

While methane fluxes are attributed to anaerobic reactions occurring primarily in the sludge layer, carbon dioxide emissions result from anaerobic, facultative, and aerobic reactions. Similarly, nitrous oxide may be formed in the soil adsorption system as an intermediate product during nitrification and denitrification processes under low oxygen conditions, with high moisture and abundant nitrogen in the forms of urea, ammonia, or proteins (Bogner et al., 2007; U.S. EPA, 2009).

The flux of methane (geometric mean, M_g, and standard deviation, s_g) measured directly from individual septic tanks (excluding Site 4) was found to range from 6.3 (s_g = 1.6) to 17.9 (s_g = 1.9) g/capita·d. Site 4 was excluded from most of the analysis because it had been pumped out two months prior to the study and was found to be producing only small amounts of gas. Further discussion about the Site 4 septic tank and its GHG emission rates is presented in Section 5.3.3.

The geometric mean of methane flux values based on all flux chamber measurements (Sites 1 through 8, excluding Site 4) is approximately 11.0 (s_g = 2.2) g/capita·d. A plot of the data for Sites 1 through 7 (excluding Site 4) and Site 8 are plotted on Figure 5-1. Site 8 was plotted separately because it is in an area with a hard water supply and the values were collected over a 24-hr period. As shown on Figure 5-1, the slope of the curve fit for the Site 8 data is steeper than that for the rest of the sites. The steeper curve fit may be an indication of greater system instability as Site 8 was found to have excess solids beyond the amount recommended for tank cleaning. Another possibility is that the sampling from Sites 1 through 7 took place during the middle of the day, while the sampling from Site 8 took place over two all day sampling events.

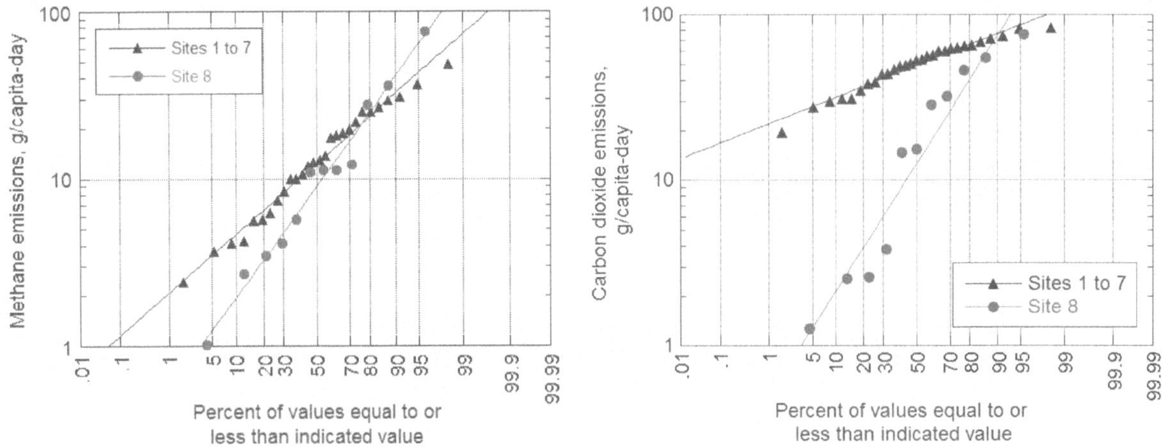

(a) (b)

Figure 5-1. Emission Rate Values Measured Using Flux Chambers in Septic Tanks for (a) Methane from Sites 1 to 7, Excluding Site 4 ($R^2 = 0.96$), and from Site 8 ($R^2 = 0.98$) and (b) Carbon Dioxide at Sites 1 to 7, Excluding Site 4 ($R^2 = 0.94$) and from Site 8 ($R^2 = 0.84$).

As mentioned previously, carbon dioxide emission rates from septic tanks can be attributed to various metabolic processes taking place in the tank, including the anaerobic degradation of organic matter in the sludge layer and facultative activity occurring in aerobic and anoxic zones throughout the tank. The geometric mean of carbon dioxide flux values based on all flux chamber measurements (Sites 1 through 8, excluding Site 4) is 33.3 ($s_g = 2.7$) g/capita·d. The flux of carbon dioxide from individual tanks was found to range from a geometric mean of 30 ($s_g = 1.4$) to 59 ($s_g = 1.3$) g/capita·d. As shown on Figure 5-2, the carbon dioxide emission rates direct from septic tanks at Sites 1 through 7 (excluding Site 4) had less variability than the methane emission rates. It was observed that the carbon dioxide emission rates from Site 8 had a different distribution than the other septic tanks, and, therefore, it was plotted separately as shown on Figure 5-2.

The mean carbon dioxide flux from the septic tank at Site 8 was about a quarter of the value measured from the other sites. One possible explanation for the low carbon dioxide flux is attributed to a reaction with calcium carbonate likely present in high levels in the water supply at Site 8, discussed further in Section 5.3.1. By comparison, Sites 1 through 7 had a relatively soft water supply.

Septic tanks were not found to be a significant source of nitrous oxide. Nitrous oxide emission rates from septic tanks were found to be negligible using the flux chambers when sampling directly from septic tanks. The measured nitrous oxide concentrations were around 0.31 ppm, which corresponds to ambient concentrations. When considering all sites, the flux of nitrous oxide ranged from 0 to 0.03 g/capita·d, with a geometric mean of 0.005 (s_g = 4.35) g/capita·d.

5.1.2 GHG Emission Rates Measured Using the Vent Method

It was found that the air movement in the household drainage system originates in the soil dispersal system and flows back through the septic tank headspace and out of the building vent. Based on this finding, it was proposed that the gas emissions from septic systems could be assessed by sampling from the vented gases in the household drainage system because the gases measured in the vent system integrate the emission rates from both the septic tank and the soil dispersal system and, therefore, may be a good representation of the overall emissions of a septic tank system.

Average emission rates of methane, carbon dioxide, and nitrous oxide, measured using the vent method (i.e., combined emission rate from septic tank and soil dispersal system), were 10.7 (s_g = 1.65), 335 (s_g = 2.1), and 0.2 (s_g = 3.6) g/capita·d, respectively. There was general agreement between the flux chamber and vent method for methane, indicating that the primary source of methane gas was the septic tank itself (see Figure 5-2). In contrast, the carbon dioxide emission rates using the vent method greatly exceeded the amount produced in the septic tank, indicating that there is significant carbon dioxide production in the soil dispersal system. A detailed comparison of the vent sampling and flux chamber methods is presented in Section 5.2.1.

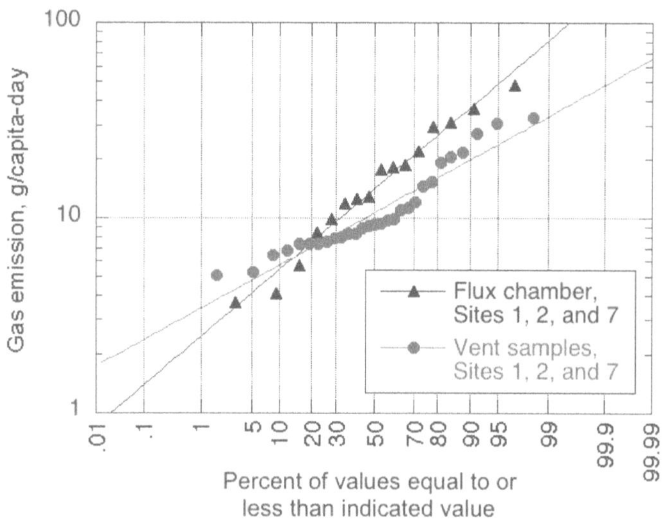

Figure 5-2. Comparison of Methane Emission Rates Using the Flux Chamber and Vent Method for Sites 1, 2, and 7.

5.1.3 Gas Concentrations in Septic Tank Liquid Samples

Gases were extracted from septic tank liquid samples to determine the amount of dissolved and entrained gases present. Nitrous oxide was not present in the liquid samples above the detection limit. The theoretical solubility values for methane, carbon dioxide and nitrous oxide in mg/L are 0.0004, 0.66 and 0.34, respectively. The methane and carbon dioxide concentrations in the liquid were measured at Sites 1, 2, and 7. The average carbon dioxide concentrations in the septic tank first and second compartment were 15.6 and 6.3 mg/L, respectively, and for methane were 4.0 and 1.3 mg/L, respectively. The measured concentrations for carbon dioxide and methane in the septic tank liquid samples at these sites were high in relation to the theoretical solubility limits. One possible explanation for the high liquid phase concentrations of these gases is that they are present as small bubbles that do not effervesce readily due to their small size.

Based on these measurements, it is estimated that the methane discharged with the effluent from the septic tank varies from 0-1.4 g/capita·d, or 0-11% of the total methane generated (flux chamber emission rate value plus gases discharged with liquid). The liquid methane values are lower than those given by Sasse (1998). Sasse (1998) suggested that dissolved methane generated in a septic tank could range from 25-50%. While Sasse (1998) does not go into detail on the origins of the percentages for dissolved methane, all the values used in the Sasse (1998) model were based on statistics from septic operated at higher temperatures and loading rates. As shown on Figure 5-3, there was not a clear correlation between the gas emission flux and the effluent aqueous methane.

At Site 8, the dissolved methane concentrations in the first and second compartment were 2.6 and 1.9 mg/L and for carbon dioxide 12.1 and 10.0 mg/L, respectively. Compared to the other sites, these dissolved methane concentrations were low in the first compartment and similar in the second compartment. The carbon dioxide concentrations had a low value in the first compartment compared to the other sites and a high value in the second compartment. Dissolved nitrous oxide was not detected in the effluent liquid at Site 8.

5.1.4 Gas Emission Rates from the Soil Dispersal System

Flux chambers placed directly above the effluent pipes in the soil dispersal system were used to estimate the GHG emissions to the atmosphere resulting from diffusion of gases through the soil. However, it was found that the GHG concentrations obtained from flux chambers located above the soil dispersal system were similar to concentrations in ambient atmosphere samples. As discussed previously, the flow of air from the soil dispersal system back through the building vent system is a likely explanation for the lack of soil-based gas emissions. It is proposed that the semi-constant negative pressure in the soil dispersal system acts to pull off-gases from metabolic processes in the soil through the effluent dispersal pipes and building vents. For example, during the first sampling event at Site 1, gas flux was found to be zero above the soil dispersal system, however, a methane emission rate of 0.8 g/capita·d was detected for the control sample. During the next sampling event at Site 1, only one of the six samples from above the soil dispersal system had measurable emission rates of carbon dioxide and nitrous oxide, however both values were similar to the control. For Site 7, carbon dioxide and nitrous oxide emission rates were measured at similar concentrations in the control and in two of the samples from above the soil dispersal system.

5.2 Mass Balance Analysis

A mass balance analysis was used to determine and compare gas emission rates from the emission sources (septic tank and soil dispersal system) identified in the previous section. The analysis presented below includes mass balances on (a) the septic tank only for a comparison of the flux chamber and vent sampling methods, (b) the septic system to determine the overall atmospheric emissions of GHG, and (c) the soil dispersal system to assess the fate of carbon. The percentage of methane and carbon dioxide in the measured emission rates is also presented in this section.

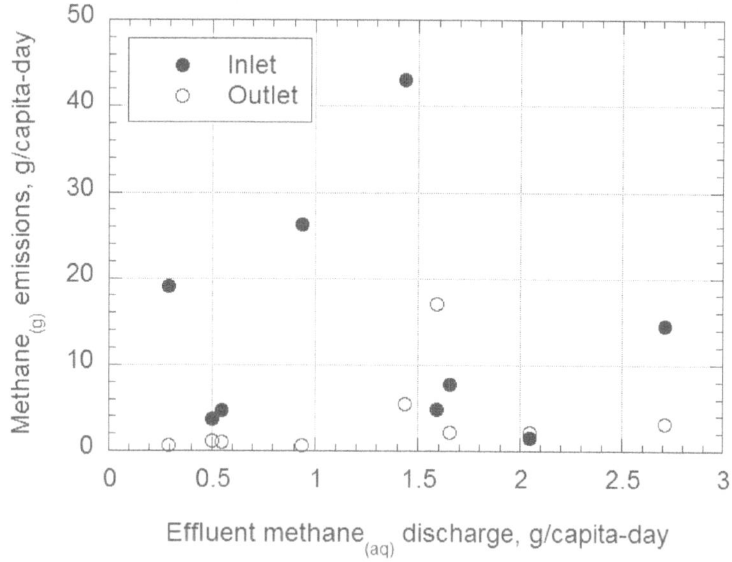

(a)

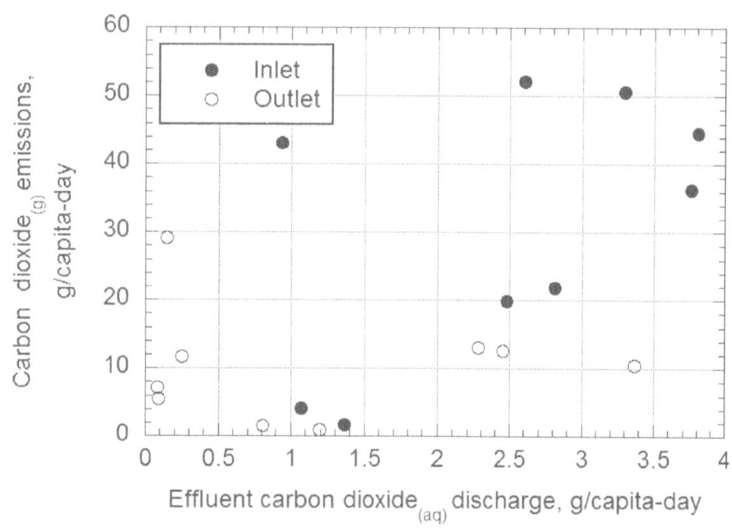

(b)

Figure 5-3. Comparison of Gas Emission Rates from Septic Tanks and from Dissolved Gases in the Septic Tank Effluent: (a) Methane and (b) Carbon Dioxide.

5.2.1 Mass Balance on the Septic Tank

A mass balance around the septic tank can be made to compare the results of the gas emission rates measured with the flux chamber and with the vent method for Site 1. Only Site 1 is used for this analysis because cleanout ports that could be used for gas sampling were located both before and after the septic tank, allowing for differentiation of the tank and soil system emission rates. As shown in Figure 5-4, the gases coming from the soil dispersal system were measured at the cleanout port located in the pipe after the septic tank (vent sample point V-1-2) and the composite gases leaving the tank (soil dispersal system + septic tank) were those measured at the cleanout located before the septic tank (Vent sample point V-1-1). The net emission from the septic tank is obtained by subtracting the gas emission rates value measured at V-1-2 from that at V-1-1. The results from the mass balance, in g/capita·d for methane, carbon dioxide and nitrous oxide are 8.4, 423.4, and 0.29, respectively. These can be compared with values of 17.9, 54.4, and zero as measured using the flux chamber in the septic tank.

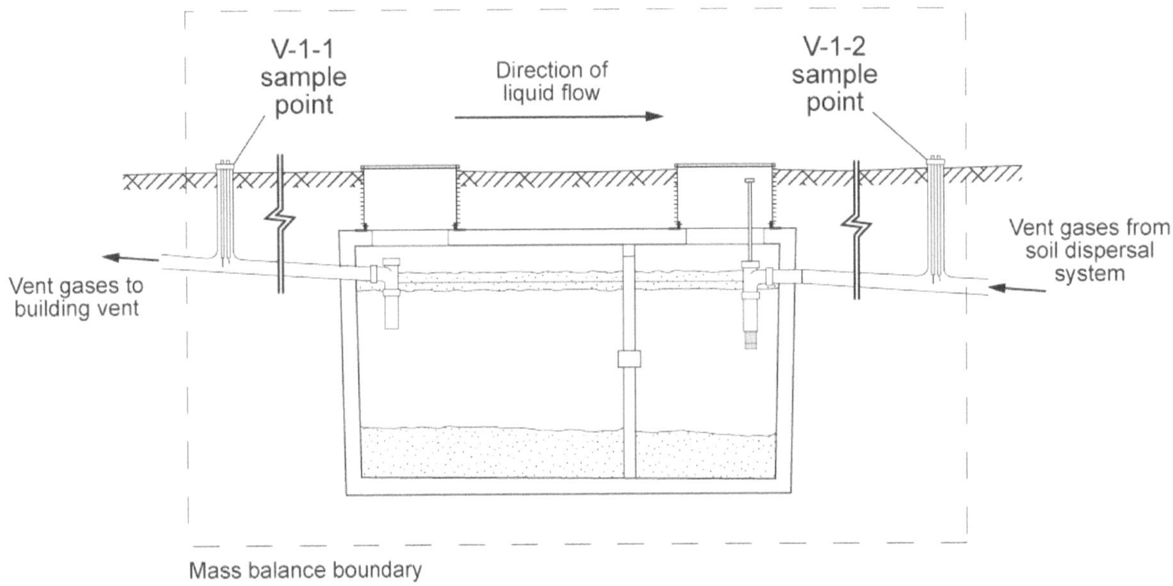

Figure 5-4. Definition Sketch for Mass Balance for Gases Moving Through the Septic Tank.

As presented in Table 5-1, the mass balance value for methane using the values measured with the flux chamber was higher than that measured with the vent method. Potential reasons for the positive bias in the flux chamber measurements compared to the vent measurements are (a) the flux chamber method draws samples from near inlet where wastewater enters the tank and possibly results in increased microbial activity, (b) wastewater discharges into the tank cause some mixing in the tank that dislodges gas bubbles from the sludge layer near the inlet, (c) the gas velocity measured in the vent system using the anemometer was lower than the actual mean velocity, and (d) insufficient samples were obtained to characterize the distribution. However, further work is necessary to determine which of these reasons (if any) is the actual cause of the discrepancy. It should be noted that if (a) or (b) is occurring, the value measured using the vent system may be more representative of the actual emission rates, whereas an incorrect velocity measurement (c) would suggest that the flux chamber measurements may be more accurate. Additional sampling should be conducted to eliminate item (d) as a possibility.

Note that methane was not detected above the ambient background in the gas samples taken at sample point V-1-2. However, a relatively high flux of carbon dioxide and nitrous oxide was determined at sample point V-1-2, resulting from the aerobic degradation of septic tank effluent in the soil. The measured results for all of the GHG's are shown in Table 5-1.

Table 5-1. Comparison between Mass Balance Values and Actual Measurements for Site 1.

	Gas emission rate, g/capita·d					
	CH_4		CO_2		N_2O	
Site	MB[a]	FC[b]	MB	FC	MB	FC
1	8.4	17.9	423.4	54.4	0.29	0.0

[a] MB = Result from the mass balance based on subtracting the emission rates measured at V-1-2 from those measured at V-1-1.
[b] FC = Value measured with the flux chamber.

A comparison of the methane emission rates obtained with the flux chamber and the vent method is shown on Figure 5-5. The vent data have less variability than the flux chamber data, as indicated by the shallower slope of the trend line fit through the data. The reduced variability from the vent system data is likely to be the result of the composite nature of the vent sample (flux values averaged over the whole system), compared to the instantaneous measurement obtained with the flux chambers (flux value extrapolated based on emission rate measured for a small area).

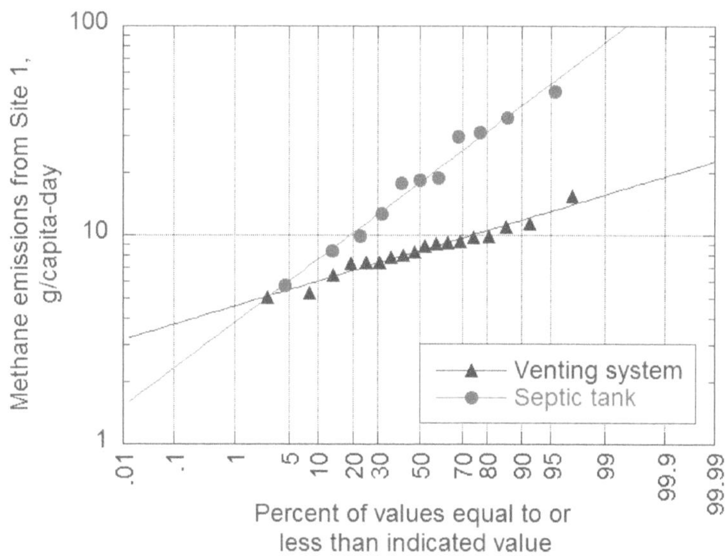

Figure 5-5. Comparison of Methane Emission Rates from the Venting System (R^2 =0.96) and the Septic Tank at Site 1 (R^2 =0.96).

5.2.2 Mass Balance on the Septic System

A mass balance on the septic system was performed to determine the overall atmospheric emissions from the system. As shown in Figure5-6 the emissions from the entire system consist of atmospheric emissions from the building vent to the air, atmospheric emissions from the soil dispersal system to the air, and gases discharged with the effluent to ground water. It should be noted that in this approach it is assumed that there are no gases escaping elsewhere in the system. Sites 1 and 7 were used for this analysis because measurements were made from both the vent and above the leach fields. In all cases the discharge of gases to the groundwater was assumed to be negligible.

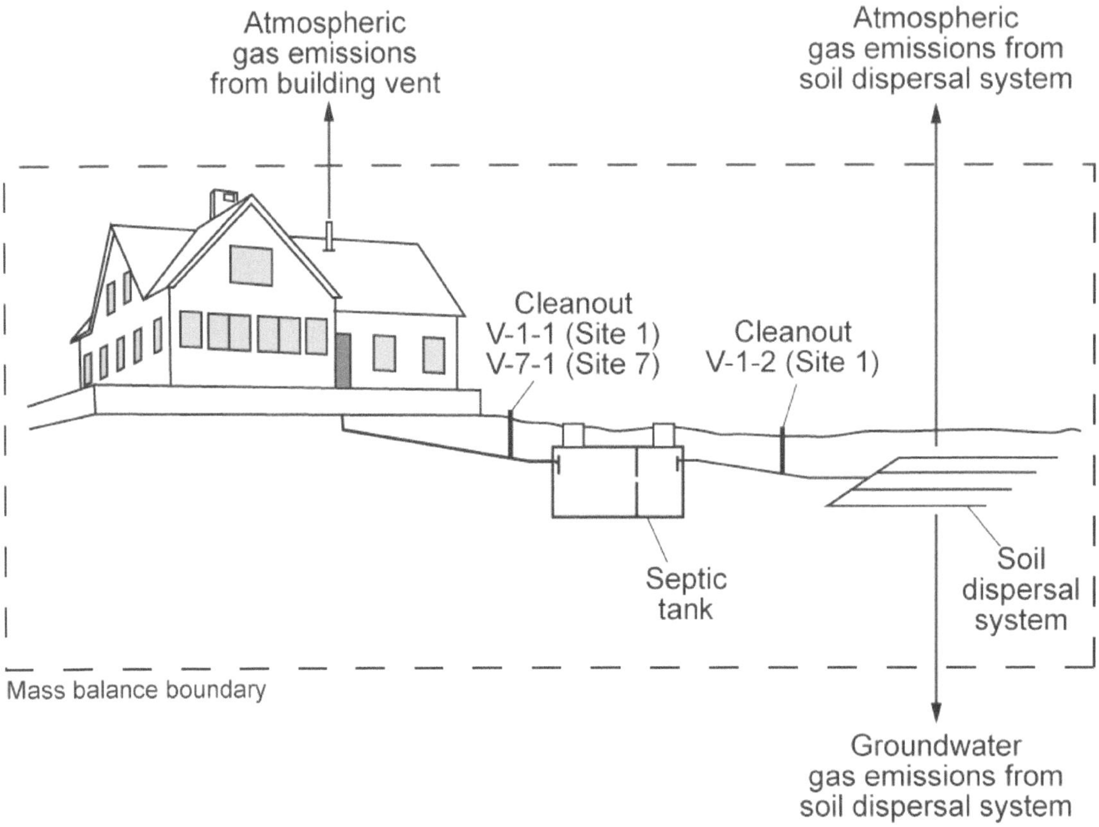

Figure 5-6. Definition Sketch for Mass Balance for Total Gas Emission Rate from the Septic System.

As discussed in Section 5.1.4, the gas emission rates from the soil dispersal system were not significantly different from the experimental control, possibly because the gases are being withdrawn through the ventilation system. Therefore, the overall atmospheric emissions from the septic system can be estimated entirely from the samples taken from the vent system located before the septic tank.

Based on the measurements at V-1-1 the overall geometric mean of the emission rates at Site 1 for methane, carbon dioxide, and nitrous oxide are 8.4, 527.0, and 0.37 g/capita·d, respectively. For Site 7, the emission rate values for methane, carbon dioxide, and nitrous oxide were 13.4, 93.0, and 0.04 g/capita·d, respectively. Note that these values are unadjusted for potential errors in the gas velocity measurement, as noted above. A summary of the mass balance results is presented in Table 5-2.

Table 5-2. Results of the Mass Balance Analysis on the Emission Rates from Septic Systems. Note: All of the Atmospheric Gas Emission Rates from These Two Systems were from the Building Vent.

Site	Gas emission rate, g/capita·day[a]		
	CH_4	CO_2	N_2O
1	8.4	527.0	0.37
7	13.4	93.5	0.04
Average	11.2	310.2	0.20

[a] The reported values correspond to averages from all the vent measurements at each site and can be found in the Appendix H.

5.2.3 Mass Balance in the Soil Dispersal System

To determine the fate of carbon present in the septic tank effluent, a mass balance analysis in the soil dispersal system was performed. The input of carbon to the system can be estimated from the COD of the septic tank effluent. The carbon dioxide equivalent of COD was determined using a stoichiometric approach where domestic wastewater is represented by the compound $C_{10}H_{19}O_3N$ (Metcalf and Eddy, 2003). The calculation of CO_2 production from COD oxidation is shown in Appendix I.

As discussed previously, direct atmospheric emissions of GHGs from the soil dispersal were determined to be insignificant. The discharge of carbon to the groundwater is assumed to be insignificant. The amount of carbon leaving the soil dispersal system can therefore be estimated by the vent samples obtained from V-1-2 as shown in Figure 5-7. Based on the system carbon balance, the carbon dioxide equivalent of the COD should be approximately equal to the CO_2 leaving the soil dispersal system through the ventilation system.

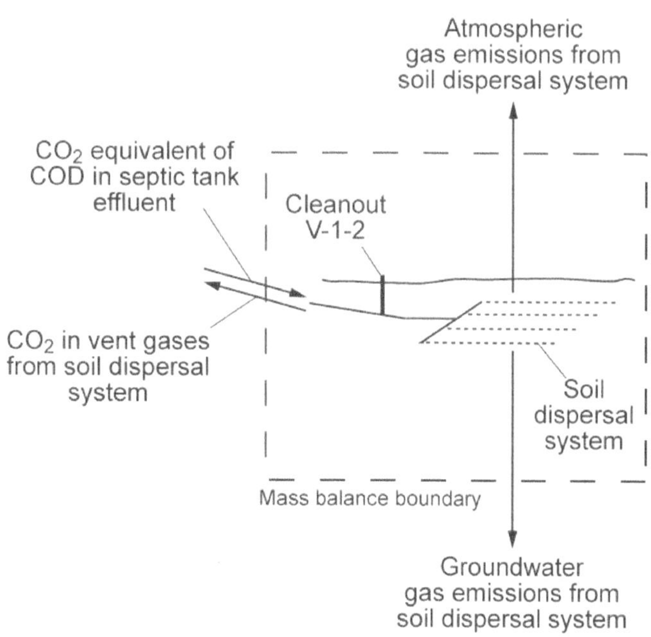

Figure 5-7. Mass Balance on the Soil Dispersal System.

The septic tank effluent CO_2 equivalent of the COD is calculated to range from 45-133 g/capita·d, depending on flowrate (see Appendix I), and the average CO_2 emission rates measured at V-1-2 is 104 g/capita·d. The approximate agreement between these values is an indication that the carbon that enters the soil dispersal system is mostly oxidized in the soil to CO_2 and this CO_2 flows back through the drainage pipes and escapes to the atmosphere through the building vent.

5.2.4 Percentage of Methane and Carbon Dioxide in the Overall Emissions

The majority of the gases expected from anaerobic degradation are methane and carbon dioxide. Therefore, the percent of each gas per sample was calculated assuming these were the only two gases present. It was found that at Sites 1 to 7 the average methane content was approximately 35% (by volume) and 65% carbon dioxide (see Figure 5-8). The high carbon dioxide percentage may indicate that there are other microbial processes occurring in the septic tank, such as aerobic or facultative bacterial activity, in addition to methanogenesis. The large amount of air moving through the venting system and headspace of the tank removes methane and carbon dioxide gases and supplies oxygen to the liquid or scum surface, potentially creating an environment for aerobic degradation to take place. Gas emission rates from Site 8 had a higher average methane content, around 65% and therefore 35% for carbon dioxide (see Figure 5-8), which agrees with the results of Winneberger (1984), where a gas mixture composition of approximately 70% methane in a septic tank was reported.

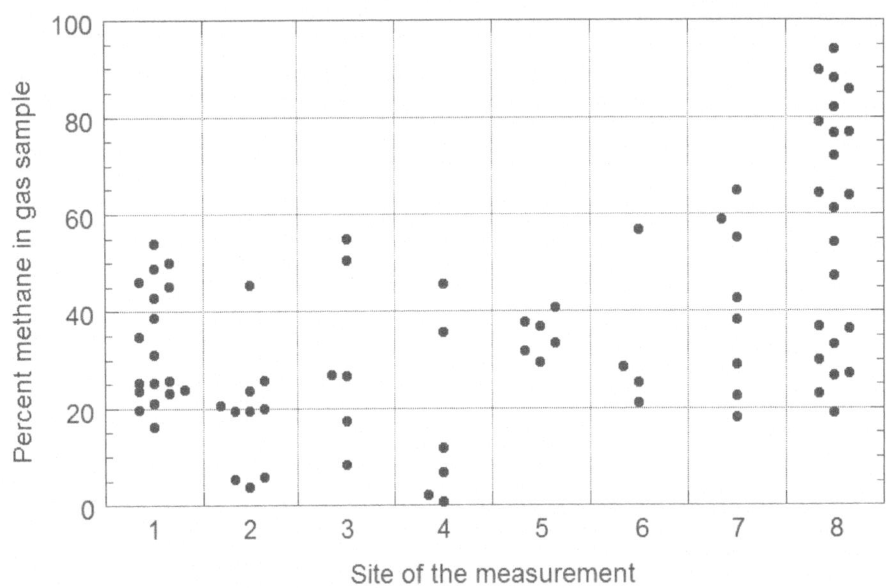

Figure 5-8. Percent Methane in the Gas Mixture at All the Sites.

The mixture that would be expected typically in an anaerobic digester is 65% methane and 35% carbon dioxide (Metcalf and Eddy, 2001). Site 8 has a similar gas mixture composition to the anaerobic digester, which may be an indication that this tank had a higher anaerobic activity than the other tanks. Another reason for the lower average carbon dioxide emission rate at Site 8 is related to the possible reaction of carbon dioxide with the high calcium carbonate content of the water supply, which is discussed in Section 5.3.1.

5.3 Sources of Variability in Gas Emissions

A number of factors contributed to the wide variability in the gas emission measurements recorded for the septic systems included in this study. The influence of factors such as water hardness, presence of scum layer, pumping intervals, turnover events, presence of invertebrates, temperature, and the septic tank effluent filter are discussed in this section.

5.3.1 Water Hardness Influence on Carbon Dioxide Emissions

As mentioned previously, the water supply for the Auburn Lakes Trails (ALT) development, encompassing Sites 1 through 7, originates in the Sierra and flows into the Strumpy Meadows Reservoir; it has an average hardness content of 9 mg/L as $CaCO_3$ (GDPUD, 2009) which indicates a soft water supply. In contrast, the water that serves the tank located in Davis is pumped out from a groundwater well and has an average hardness of 212 mg/L as $CaCO_3$ (CDPW, 2009) indicating a hard water supply.

A comparison of the carbon dioxide emission rates data between the sites located in ALT and the tank for Site 8 (Figure 5-2) indicates different distributions. It is hypothesized that the hardness content of the water supply in each area may be a reason for this differing behavior. In addition to the difference in source water, Site 8 had other distinguishing characteristics that potentially impacted emissions. Unique aspects at Site 8 include the diversion of laundry water to a gray water system, a higher number of occupants per tank volume, and the vegetarian diet of the house occupants. It is conceivable that these aspects may also influence the overall anaerobic and facultative processes occurring in the tank and hence the carbon dioxide emission rates.

5.3.2 Influence of Scum Layer on Gas Emissions

Two different groups of sites were compared to determine the influence of the scum layer in the overall emissions from the septic tank. The first group was composed of the tanks from Sites 2, 3, and 7 that share as a common characteristic a thick scum layer varying from 3 to 5 inch in depth, with similar appearance, black color, humus like texture and usually covered the liquid surface of the septic tank. The septic tanks at Sites 1, 5, and 6 formed the second group; these tanks were characterized by patchy, light scum (less than 1"), with a light brownish color. As shown on Figure 5-9, the average methane and carbon dioxide emission rates from the sites that have a thick scum layer are similar to the sites without scum and the slopes are also similar. The data from the septic tank at Site 8 (light scum, hard water) was plotted separately for comparison.

5.3.3 Emissions from Recently Cleaned Tank

The septic tank located at Site 4 was not a typical septic tank. It did not have any scum formation; the sludge was less than 14" in depth, with a light brown coloration and a particular pine odor was emitted when the tank lids were opened. The tank was last pumped out in late July of 2009 (measurements took place during September 2009). The methane and carbon dioxide gas emission rates from the tank at Site 4 in g/capita·d, averaged 0.2 and 3.2, respectively. Nitrous oxide emission rates were negligible. From the results of the overall GHG emission rate and the characteristics of the tank discussed above, it can be assumed that methanogenesis was not occurring in this tank, which is consistent with results reported by Weibel et al. (1955) for recently pumped septic tanks, which had a lag phase in the gas production.

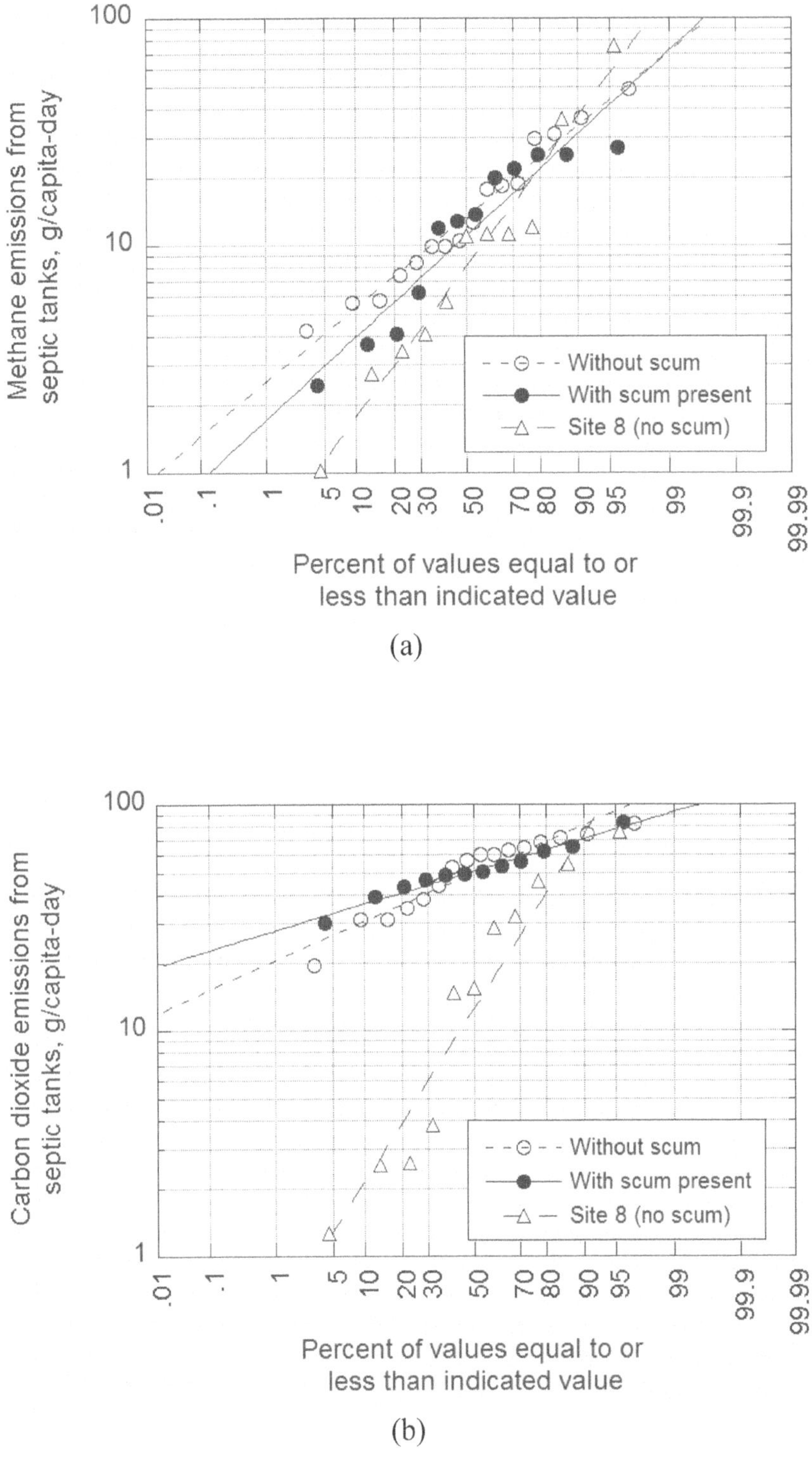

Figure 5-9. Gas Emission Rates Results Grouped Based on Presence of Scum: (a) Methane and (b) Carbon Dioxide.

5.3.4 Influence of Turnover Event on Septic Tank Gas Emissions

Several turnover episodes were observed in the septic tank at Site 1 during the course of the study. As shown in Figure 5-10, during the turnover event, a large number of bubbles and sludge came to the surface, causing the tank contents to mix with subsequent gas release to the atmosphere. A gas sample taken using the flux chamber before the turnover event had a flux of methane of 43 g/capita·d while the methane flux after the turnover event was 18 g/capita·d. The reduction in the emission rates indicates that the tank released the methane gas trapped in the sludge during the turnover events. It was surprising to observe these episodes during fall and early winter in light of the observations of Winneberger (1984) and D'Amato et al. (2008) that these events (will) happen more frequently during the spring and summer months when anaerobic activity increases due to the rise of temperature..

Figure 5-10. Views of Tank at Site 1 (a) Just Before and (b) During a Turnover Event.

5.3.5 Variability in Daily GHG Emissions

High variability was observed in the methane emission rates from tank to tank and from sample to sample for a given tank, highlighting the danger of generalizing on the basis of single measurements or single sites. A good example is Site 8, where the 24-hour sampling events took place. As shown in Figure 5-11, two particular periods with comparably high methane emission rates were observed, one at 7:00 am and another at 11:00 pm.

These high emission periods correlated with activities that were taking place in the house at the time of sampling, bathing during the 7:00 am sample and manual dishwashing at the 11:00 pm sample, indicating that the high methane emission rate values may be related to the water usage in the house. It is proposed that the sludge at the bottom of the tank may be disturbed when water is discharged to the septic tank, resulting in the release of gas bubbles that have accumulated in the sludge layer. The flux of carbon dioxide does not appear to follow the same trend as methane, which may be related to the reactions of carbon dioxide in the water. The cause of the low emission rate of carbon dioxide on the 12/10/09 sample event has not been determined.

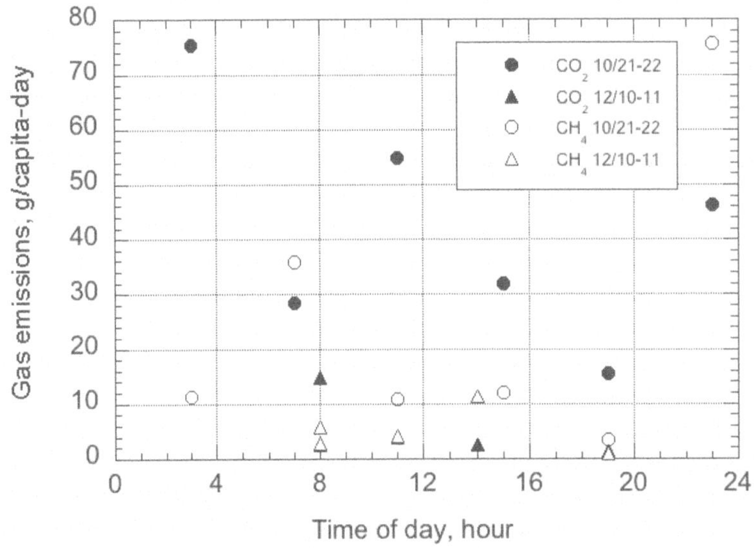

Figure 5-11. Emission Rates from Site 8 During the Two 24-Hour Sampling Events for Methane and Carbon Dioxide.

5.3.6 Relation between Invertebrates and Scum Layer

Sites 2, 3 and 7 had a thick scum layer and the presence of invertebrates such as red worms and small flies. It was confirmed that a thick scum layer with dark brown to black coloration and with the appearance of "crumbly earth" or humus were populated by earthworms as reported by Winneberger (1984). It was observed at the time that small flies and its larva were also living on the upper layers of the scum, as previously reported by Dale (1982). However, a larger number of organisms were not found to correspond to a thicker layer of scum as was observed by Dale (1982). Small black larvae were observed at Sites 6 and 8 in where the scum layer was thin (less than 3 cm).

5.3.7 Temperature Influence on Gas Emissions

The average liquid temperature inside the tanks ranged from 12-27°C during the four months of sampling. Kinnicutt et al. (1910) and Winneberger (1986) reported a reduction in gas emissions during cold months, however, no correlation between the average liquid temperature inside the septic tanks and the overall methane and carbon dioxide emissions was observed in this study (Figure 5-12). The sites with the largest number of measurements (Sites 1 and 8) were also plotted separately to identify a possible trend with temperature; none is apparent (Figure 5-12 c and d). While it is possible that the sampling duration and range of temperatures in this study were insufficient to characterize the seasonal temperature variation, the more likely reason is that factors other than temperature play a more important role in determining the measured emission rates at any given time/location (for example, the diurnal measurements presented on Figure 5-12).

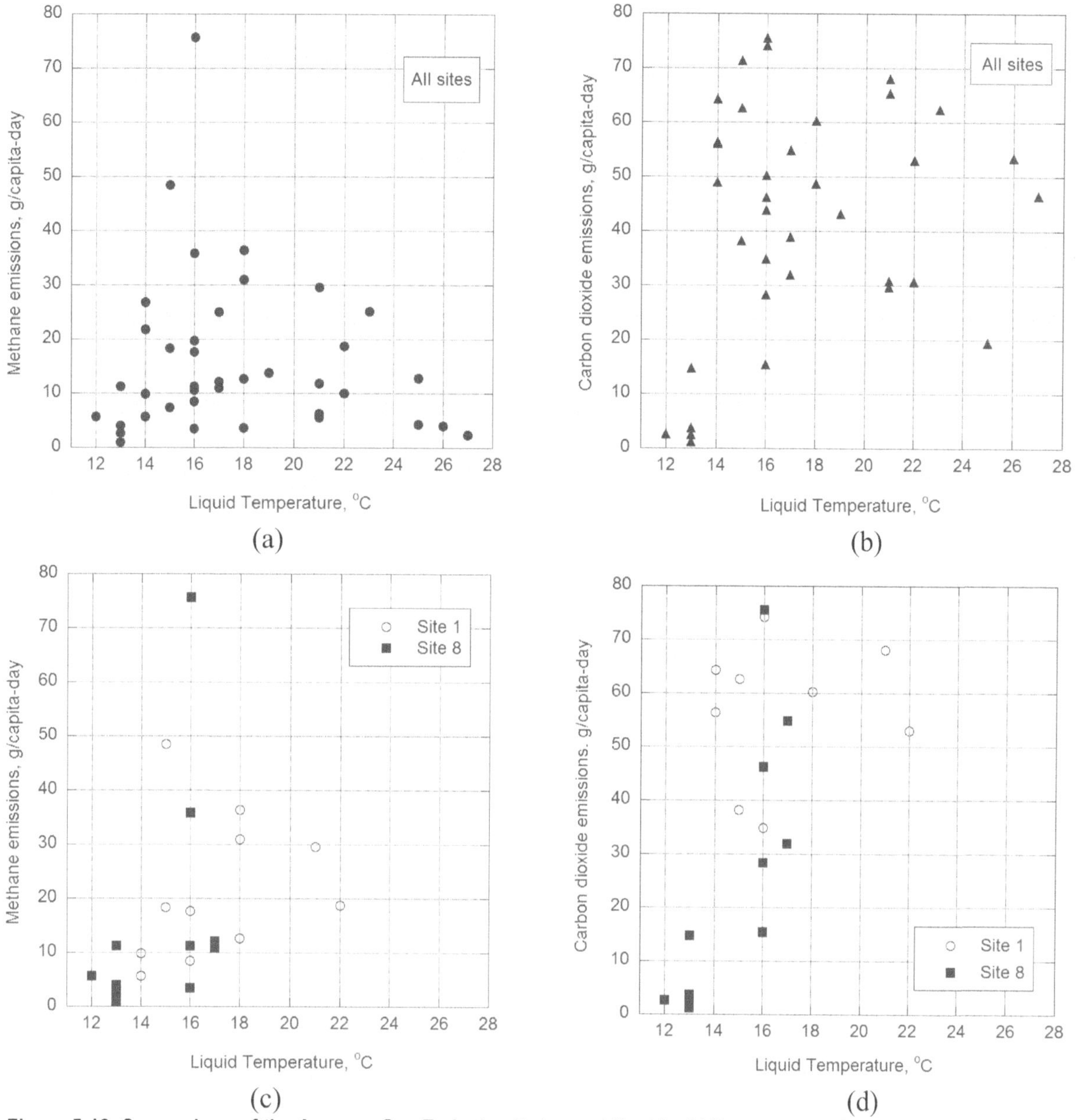

Figure 5-12. Comparison of the Average Gas Emission Rates and the Liquid Temperature: (a) Methane and (b) Carbon Dioxide Emission Rates at Sites 1 to 7, (c) Methane and (d) and Carbon Dioxide Emission Rates at Sites 1 and 8.

5.3.8 Influence of the Effluent Filter in the Venting System

The influence of the effluent filter on the air flow from the septic tank through the vent system was evaluated at Sites 1 and 7. It was found that gas flow rates in the passive house venting systems ranged from 150 to 400 m^3/d when an effluent filter was not present. Effluent filters reduced the air flow through the vent system to a range of 10 to 70 m^3/d, depending on the type of filter (see Figure 5-13). Thus, it was found that effluent filters may obstruct the natural flow of air through the tank headspace. Flow of air through the tank headspace is important for the evacuation of methane (an explosive gas) and odor causing compounds. Thus, improved designs for effluent filters may be needed to prevent the accumulation of these gases in the tank headspace, particularly under calm weather conditions when flow in the vent systems is reduced.

(a)

(b)

Figure 5-13. Views of the Effluent Filters: (a) Filter in Effluent Tee and (b) Comparison of Two Types of Filters. Internal Elements of the Filter are Removed for Viewing.

CHAPTER 6.0

IMPLICATION OF RESULTS

An important part of this research is to provide field data that can be used as a basis for comparison with previous studies and for calibration of GHG emission inventory models. In the following discussion, the emissions findings from this study are: 1) compared to values found in the literature review and model values and 2) evaluated in terms of their global warming potential.

6.1 Comparison of Gas Emissions to Literature Values and Models

A summary of the literature emission values and those measured using the flux chamber is presented in Table 6-1. The average methane emission rate obtained using flux chambers in the septic tanks were in general agreement with the estimates of methane emission rates based on the COD loading and Kinnicutt et al. (1910) models. However, these averages are somewhat lower than the estimates from Winneberger (1984) and Sasse (1998) and much lower than the IPCC (2007) model for methane emission rate in septic tanks.

Table 6-1. Summary of Methane Emission Rates, Including this Study.

Method	Year	Methane estimate ($g\ CH_4$/capita·d)
Kinnicutt et al.	1910	10.1[a]
Winneberger	1984	14 to 18[a]
COD loading	2010	11[b]
IPCC	2007	25.5[c]
Sasse	1998	18[d]
This study	2010	11.0 (s_g = 2.50)[a,e]

[a] Measured value from community septic tanks.
[b] Calculated value assuming that 40% of solids are removed as septage (see Appendix B).
[c] Calculated value assuming that half of the influent BOD is converted anaerobically (see Appendix C).
[d] Calculated value assuming 25% CH_4 dissolved (see Appendix D).
[e] Geometric mean and standard deviation as determined using flux chamber method, this study.

One reason for the difference between the measurements presented here and those determined from the IPCC (2007) model is that the IPCC (2007) model assumes that half of the influent BOD to the septic tank is converted anaerobically. Further, the fate of organic matter present in septage and septic tank effluent is not accounted for clearly. It is likely that the measurements made by Winneberger (1984) are high because not enough samples were obtained to establish a statistical distribution. The values used in the Sasse (1998) model were based on statistics from septic tanks located in developing countries, which may operate at higher temperatures and loading rates, resulting in higher methane emission rates.

It is important to note that the U.S. EPA (2010) GHG emissions model for septic systems uses an approach similar to the IPCC (2007) model; using a methane correction factor (MCF) of 0.5 (see Section 2.5.2). However, this value does not specifically account for the fate of the sludge when the tank is cleaned out or for effluent BOD oxidation in the soil dispersal system. Based on the mean methane emission value measured in this study (11 g CH_4/capita·d), an MCF value of about 0.22 would be applicable for the systems evaluated in this research project.

6.2 CO_2 Equivalent of Gas Emissions Values

The septic tank emission rates for methane and nitrous oxide have been converted to CO_2-equivalent (CO_2e) emissions using the global warming potential for a 100 year horizon based on the IPCC (1996) values. The CO_2e is calculated by multiplying the measured emission rates by the GWP, as summarized in Table 6-2.

Table 6-2. Summary of Emission Rates as CO_2e from the Septic Tank and Vent Average Measurements.

	Geometric mean emission rate value (g/capita·d)			Carbon dioxide equivalent emissions (tonne CO_2e/capita·year)	
Compound	Septic tank	Septic system[b]	GWP[a]	Septic tank	Septic system[b]
Methane	11.0	10.7	21	0.084	0.082
Nitrous oxide	0.005	0.20	310	0.00057	0.023
Carbon dioxide	33.3	335	1	0.012	0.12
Total GHG emissions				0.096	0.23
Total anthropogenic emissions[c]				0.085	0.10

[a] GWP for a 100 year horizon IPCC (1996).

[b] As determined from vent system sampling.

[c] Biogenic carbon dioxide is not included in GHG inventories (U.S. EPA, 2009).

As shown in Table 6-2, methane generation from the septic tank is the primary source of anthropogenic GHG emissions. The CO_2e GHG emission rates from septic tank systems determined using either the flux chamber or mass balance methods are relatively low compared to those for a citizen of an industrialized country (about 23 tonne CO_2e/capita·year; U.S. EPA, 2010). Using the total emission values reported in U.S. EPA (2010), the per capita GHG emissions associated with wastewater treatment are 0.13 and 0.92% for centralized and decentralized (i.e., systems with onsite septic tanks), respectively. However, using the adjusted emission values as determined in this study, the GHG emissions associated with septic tank type systems are about 0.47% of the average per capita GHG emissions, resulting in an estimated total emission of about 6.5 Tg CO_2e/year from septic tank systems. It should be noted that these GHG inventories do not account for the embodied GHG emissions associated with construction or maintenance of infrastructure, process chemical and energy inputs, or downstream environmental impacts. Thus, this type of GHG emissions comparison is limited in that it is based solely on fugitive methane and nitrous oxide emissions.

CHAPTER 7.0

FINDINGS AND RECOMMENDATIONS

The major findings of this research study, other findings, and recommendations for future studies are presented in this section.

7.1 Major Findings

The principal findings from this research are:

- The geometric mean of the total emission rates for methane, carbon dioxide, and nitrous oxide based on samples from the vent system were 10.7 ($s_g = 1.65$), 335 ($s_g = 2.13$), and 0.20 ($s_g = 3.62$) g/capita·d, respectively.
- The CO_2e of the methane emission rate to the atmosphere, as determined from the vent system data in this study, is about 0.10 tonne CO_2e/capita·yr. Based on the current estimated per capita CO_2e emission rate for the United States (i.e., 23 tonne CO_2e/capita·yr; U.S. EPA, 2010), the septic tank system accounts for about 0.5% of the total per capita emission.
- The septic tank methane flux values determined by the flux chamber method were found to have a log-normal distribution, with a geometric mean (M_g) of 11.0 g/capita·d and geometric standard deviation (s_g) of 2.50. Similarly, the values of M_g for carbon dioxide and nitrous oxide were 33.3 ($s_g = 2.73$) and 0.005 ($s_g = 4.35$) g/capita·d, respectively.

7.2 Other Findings

Other findings from this research are:

- There was general agreement for methane gas emission rates determined with theoretical models and measured with the flux chamber and vent methods.
- There is considerable variability in the methane gas fluxes from tank to tank and sample to sample.
- The hardness of the water supply appears to influence the overall flux of carbon dioxide, with soft water systems having higher carbon dioxide gas fluxes.
- A correlation between the GHG emission rates from the septic tank and the liquid temperature was not observed, perhaps as a result of the limited duration of the study.
- The presence or absence of a scum layer had no discernable influence on emission rates of methane and carbon dioxide.
- The carbon (measured as COD) in the septic tank effluent discharged to the soil dispersal system was approximately equal to the amount of carbon being vented (measured as carbon dioxide) from the soil dispersal system.

- Methane generated during the anaerobic digestion process was found in both gaseous and aqueous forms but no relation was found between the gas and liquid phase concentrations.
- Nitrous oxide emission rates from septic tanks were found to be negligible; however nitrous oxide was detected in the gases vented from the soil dispersal system.
- Essentially no gas emissions from the soil surface were measured using flux chambers placed above the soil dispersal system trenches.
- Using the mean methane emission value measured for the septic tanks from this project, an MCF value of 0.22 was calculated.

7.3 Recommendations

Based on the findings of this research, the following recommendations are proposed for future studies of greenhouse gas emissions from septic tank systems:

- Due to the uncertainty in several operational parameters, such as temperature and water hardness, and their influence on the production of septic tank gases, further study in other regions of the country is recommended.
- In this study, only direct GHG emissions from operational septic systems were evaluated. A follow up study quantifying the GHG emissions associated with various septage management practices is needed.
- The soil dispersal systems in this study were well drained and did not have any ponding. The evaluation of soil dispersal systems in other soil types and at different stages of ponding is needed. In addition, studies on alternative soil dispersal systems, such as pressure and drip irrigation systems, are needed.
- Only conventional septic tank systems were evaluated in this study. Additional research is needed to quantify gas emissions from alternative onsite wastewater systems, such as natural treatment systems, packed bed filters, and other aerobic treatment processes.
- A study is needed to evaluate the development of methanogenesis in septic tanks from the time of start-up, with and without inoculation.
- The methods for sampling of gases from ventilation systems should be further developed and refined.
- Further study is needed to develop technologies for the control of GHG emissions from wastewater systems is needed, including soil-based filtration and stand-alone biofiltration processes. Integration of these types of control systems may require slight modifications to the aspects of the building code related to septic tank ventilation systems.
- A study is needed to determine the correlation between the GHG emissions and the septic tank influent quality and loading.
- Further work should be conducted to quantify GHG emissions from all types of wastewater management systems so that accurate models can be developed.

APPENDIX A

COD LOADING PER CUBIC METER CALCULATION IN SEPTIC TANKS

The COD loading into a septic tank was calculated assuming two people per residence and two different tank sizes: 1,000 and 1,500 gallons.

1. Determine the total COD loading per day
   The quantity of COD discharged per person is assumed to be 200 g/capita·d (Crites and Tchobanoglous, 1998):
   Total COD loading is 400 g/d (for two people)
2. Assuming a tank volume of 1,000 gallons
   (400 g/d) / (3.78 m^3) is equal to 106 g COD/m^3·d
3. Assuming a tank volume of 1,500 gallons
   (400 g/d) / (5.68 m^3) is equal to 70 g COD/m^3·d
4. COD loading ranges between 70 and 106 g/m^3·d

It should be noted that in the above computation the 200 g/capita·d represents the total COD and it is not differentiated between soluble and particulate COD. Particulate COD is expected to primarily settle out in the septic tank while dissolved COD will primarily leave the septic tank with the effluent flow. The referenced values used in the computations presented in Appendixes A and B are based on typical septic tank systems.

APPENDIX B

ESTIMATE OF METHANE EMISSIONS FROM SEPTIC TANK SYSTEM BASED ON VSS AND COD LOADING

Determine the theoretical amount of methane released per day per person discharging wastewater to a septic tank, which is pumped on a 6 yr interval. Note that for a new tank, there may be a lag in methane production, for example methane production may not occur until spring if the tank is pumped in the winter. Also, some people recommend leaving some sludge in the tank for inoculation to promote methanogenesis.

A. Based on VSS Loading

1. Determine the total VSS loading per day
   The quantity of VSS discharged per person is 35 g/capita·d (Crites and Tchobanoglous, 1998). The total VSS loading over the 6 yr operational period is 154 kg VSS for 2 people.
2. Determine the effluent VSS discharged from the septic tank
   Effluent VSS can be estimated from the effluent TSS. Typical effluent TSS is 85 mg/L (Crites and Tchobanoglous, 1998), and assuming an effluent VSS to TSS ratio of 0.5, the effluent VSS is 42.5 mg/L. The total VSS loading in the effluent over the 6 yr operational period is 41 kg VSS for 2 people.
3. Estimate the total VSS removed as septage after 6 yr (U.S. EPA, 1994)
   Assume the septage removed has a VSS of 9 g/L and a volume of 4,000 L
   The VSS removed as septage is equal to 36 kg VSS after the 6 yr operational period.
4. Estimate the VSS converted to gas
   The VSS converted to gas is determined by subtracting the effluent VSS loading (41 kg) and the septage VSS (36 kg) remaining in the tank from the influent VSS daily loading (154 kg). The resulting VSS converted to gas is 154 kg – 41 kg – 36 kg = 77 kg (over the 6 yr operational period). Thus, the overall VSS destruction rate in the septic tank is estimated to be 50 percent.
5. Determine the methane produced each day per capita
   Assuming gas production rates of 1 m^3/kg VSS converted, gas composition of 60 percent methane (Tchobanoglous et al., 2003), the methane emission rate is determined to be:
   Methane production = [(77 kg VSS x 1,000 g/kg x 0.6) / (6 yr x 365 d/yr x 2 capita)]
   = 10.6 g/capita·d

B. Based on COD Loading

1. Determine the total COD loading per day
   The quantity of COD discharged per person is 200 g/capita·d (Crites and Tchobanoglous, 1998):
2. Determine the amount of COD retained in septic tank per day
   Assume 60 percent of the influent COD is retained in the septic tank
   The COD remaining in the tank each day is equal to 120 g/d (200 g/d x 0.6)
3. Estimate the total COD removed as septage after 6 yr
   Assume the septage removed has a COD of 40 g/L (U.S. EPA, 1994) and volume of 4,000 L
   The COD removed as septage is equal to 160,000 g (40 g/L x 4,000 L)
4. Convert total COD in septage to equivalent daily value
   For 6 yr time period, the daily COD value in the septage is equal to 73 g/d [160,000 g/(6 yr x 365 d/yr)]
5. Estimate the COD converted to methane (CH_4)
   The COD converted to gas is the difference of daily loading (120 g/d) and the daily equivalent of COD of the removed septage (73 g/d)
   The COD converted to gas is 47 g/d (120g/d – 73 g/d)
6. Conversion of COD to methane at standard conditions (Tchobanoglous et al., 2003).
   COD of CH_4 is the amount of O_2 needed to oxidize CH_4 to CO_2 and H_2O

   $CH_4 + 2O_2 = CO_2 + H_2O$

   $2(32g\ O_2/mole) = 64\ g\ O_2/mole\ CH_4$
   The volume per mole of CH_4 at standard conditions is 22.4 L. Therefore, the CH_4 equivalent of COD converted under anaerobic conditions is (22.4 L/64 g)
   The conversion of COD to methane is equal to 0.35 L CH_4/ g COD
7. Determine the volume of gas produced each day per capita
   Assume the following conversion factors apply
   Density of CH_4 is equal to 0.67 g/L (Density at 20 °C)
   Methane volume produced is equal to 16.45 L/d [(47 g COD/d)·(0.35 g/L CH_4/ g COD)]
   Mass of CH_4 is 11.0 g/d [(16.45 L/d) (0.67 g/L)]
   The contribution of CH_4 per person is equal to 11.0 g/capita·d

APPENDIX C

ESTIMATE OF METHANE EMISSIONS FROM SEPTIC TANK SYSTEM BASED ON IPCC METHODOLOGY (1996, 2007)

The following estimate was based on the methodology developed in the IPCC guidelines for National Greenhouse Gas Inventories, Chapter 6, Methane Emission from Wastewater. Calculations will be based on per-capita basis.

1. Estimate organically degradable carbon in wastewater (TOW)
   TOW = P x BOD x I X 365, where:
   Where TOW = total organics in wastewater in kg BOD/yr
   P = population
   BOD = U.S. per capita BOD in inventory year (value taken from Table 6.4, Estimated BOD_5 values in domestic wastewater for selected regions and countries. The values are an assessment of the literature. Chapter 6, IPCC, 2006)
   I = correction factor, default value 1.0
   TOW = 1 person x 85 g BOD/capita·d x 1.0 x 0.001 kg/g x 365 d/yr
   TOW = 31.03 kg BOD/capita·yr
2. Calculate the methane CH_4 emission factor (EF_j)
   $EF_j = B_o$ x MCF_j , in kg CH_4/kg BOD
   Where B_o = maximum CH_4 producing capacity in kg CH_4/kg BOD (value taken from Table 6.2, Default maximum methane producing capacity for domestic wastewater. Chapter 6, IPCC 2006)
   MCF_j = methane correction factor for septic tanks (value taken from Table 6.3, Default methane conversion values for domestic wastewater. Chapter 6, IPCC 2006)
   EF_j = 0.6 kg CH_4/kg BOD x 0.5
   EF_j = 0.3 kg CH_4/kg BOD
3. Estimate the per capita CH_4 emission
   Methane emission = TOW x EF_j in g CH_4/capita·d
   CH_4 = (31.03 kg BOD/ capita·yr) (0.3 kg CH_4 /kg BOD)
   9.3 kg CH_4/capita·yr = 25.5 g CH_4/capita·d

Comment:
It should be noted that in the above computation it is assumed that half of all influent BOD is converted anaerobically.

APPENDIX D

ESTIMATE OF METHANE EMISSIONS FROM SEPTIC TANK SYSTEM BASED ON SASSE (1998)

The model presented below was developed by Sasse (1998) based on observations of septic tank operation primarily from developing countries. The model parameters, including BOD_5, water consumption per capita, COD/BOD ratio, septic tank dimensions, and dissolved methane content in water were adjusted to be consistent with typical septic tank design parameters in the U.S.

1. Table 22 from Sasse (1998), wastewater production per capita. The highlighted values in the table were calculated using the following assumptions:
   - Population: 1
   - BOD: 85 g/capita·d (Table 4.12, Crites and Tchobanoglous, 1998)
   - Water consumption: 300 L/capita (Average water used in individual residences, Crites and Tchobanoglous, 1998)
   - COD/BOD: 2.33 (COD and BOD values taken from Table 4.12, Crites and Tchobanoglous, 1998)

Table D-1. Wastewater Production Per Capita.

User	BOD_5 per user	Water consumption per user	COD /BOD_5 ratio	Daily flow of wastewater	BOD_5 conc.	COD conc.
(1)	(2)	(3)	(4)	(5)	(6)	(7)
Given	Given	Given	Given	Calc.	Calc.	Approx.
Number	g/d	L/d	mg/L / mg/L	m^3/d	mg/L	mg/L
1	85	300	2.33	**0.3**	**283**	**660**

1.1 Calculation of daily flow of wastewater, m^3/d – Column (5)
Daily Flow of wastewater = number of users x water consumption per user
Daily Flow of wastewater = 1 x 300 L/d x 10^{-3} m^3/L = 0.3 m^3/d

1.2 Calculation of BOD_5 concentration, mg/L – Column (6)
BOD_5 = BOD_5 / Daily Flow of wastewater
BOD_5 = (85 g/d) / (0.3 m^3/d) = 283 mg/L

1.3 Calculation of the approximate COD concentration, mg/L – Column (7)
COD = COD / BOD_5 x BOD_5
COD = 2.33 x 283 mg/L = 660 mg/L

2. First row of Table 23 from Sasse (1998), general spread sheet for septic tank, input and treatment data. The highlighted values on the table were calculated using the previous values from Step 1 and assuming 12 hours as the time of most wastewater flow and SS/COD ratio equal to 0.42 mg/L.

Table D-2a. General Spread Sheet for Septic Tank, Input and Treatment Data.

Daily waste water flow	Time of most waste water flow	Max flow at peak hours	COD inflow	BOD_5 inflow
(1)	(2)	(3)	(4)	(5)
Given	Given	Calc.	Given	Given
m^3/d	h	m^3/h	mg/L	mg/L
0.3	12	**0.025**	660	283
			COD/BOD_5	2.33

Table D-2b. General Spread Sheet for Septic Tank, Input and Treatment Data (Continued).

HRT inside tank	Settleable SS/COD ratio	COD removal	COD outflow	BOD_5 outflow
(6)	(7)	(8)	(9)	(10)
Chosen	Given	Calc.	Calc.	Calc.
h	mg/L / mg/L	%	mg/L	mg/L
24	0.42	**36%**	**421**	**175**

2.1 Calculation of maximum flow at peak hours, m^3/h – Column (3)
Maximum daily flow = Daily wastewater flow, m^3/d / time of most wastewater flow, hr/d
Maximum daily flow = 0.3 m^3/d / 12 h/d = 0.025 m^3/h

2.2 Calculation of the COD removal, % – Column (8)
To calculate the COD removal, Sasse (1998) propose a factor of 0.6. This factor takes into account that in a septic tank the COD removal rate depends on the amount of settleable solids, their COD content, and the intensity of inoculation of fresh inflow. The COD removal is calculated based on the chosen HRT (24 hr).
COD removal = (SS / COD) / 0.6 x {[(HRT – 3) x (0.15 / 27)] + 0.4}
COD removal = 0.42 / 0.6 x {[24 - 3] x (0.15 / 27)] + 0.4} = 36%

2.3 Calculation of the COD outflow, mg/L – Column (9)
COD outflow = (1 – COD removal rate) x COD inflow
COD outflow = (1 -0.36) x 660 mg/L = 421 mg/L

2.4 Calculation of the BOD removal, %
The equation to calculate the BOD removal is related to Fig. 65 in Sasse (1998), the simplified curve of change in the COD/BOD ratio during anaerobic treatment. The BOD removal is based on the COD removal rate (36%).
For a COD removal less than 0.5, the COD/BOD removal ratio is 1.06
Therefore, BOD removal is (0.36)(1.06) = 0.38 or 38%

2.5 Calculation of the BOD outflow, mg/L – Column (10)
BOD outflow = [1 – (BOD removal)] x BOD
BOD outflow = [1 – (0.38)] x 283 mg/L = 175 mg/L

3. Second row of Table 23 from Sasse (1998), dimensions of septic tank. The highlighted values on the table were calculated using the following assumptions:

- Desludging interval: 72 months, corresponding to 6 yr pump out interval
- Typical septic tank dimensions:
  - Inner width of 1.25 m
  - Inner length of first chamber of 1.5 m
  - Water depth at outlet point of 1.25 m
  - Length of second chamber of 1 m
- Dissolved methane: 25 percent

Table D-3a. Dimensions of Septic Tank.

Desludging interval	Inner width of septic tank	Minimum water depth at outlet point	inner length of first chamber	
(1)	(2)	(3)	(4)	(5)
Chosen	Chosen	Chosen	Required	Chosen
mo	m	m	m	m
72	1.25	1.25	0.48	1.5

Table D-3b. Dimensions of Septic Tank (Continued) and Biogas Production.

Length of second chamber		Volume including sludge	Actual volume of septic tank	Biogas 70% CH_4 (25% dissolved)
(6)	(7)	(8)	(9)	(10)
Required	Chosen	Required	Check	Calc.
m	m	m^3	m^3	m^3/d
0.24	1	**1.13**	**3.90**	**0.03**
Sludge L/g BOD rem -->		**0.0021**		

3.1 Calculation of sludge volume (Sasse, 1998), sludge L/g BOD remaining
The sludge removal BOD depends on the desludging interval, for this case 72 months.
Sludge removed as BOD = if desludging interval < 120 → 0.005 x 0.5 - [(desludging interval - 36) x 0.002]
Sludge volume= 0.005 x 0.5 – [(72-36) x 0.002] = 0.0021 sludge L/g BOD remaining

3.2 Tank volume required including sludge storage, m^3 – Column (8)
Volume required including sludge storage = 2 x daily flow, m^3/d x HRT, h / 24 h/d x inner width of septic tank chosen, m x inner length of first chamber chosen, m
Volume including sludge = 2 x 0.3 m^3/d x 24 hr / 24 hr/d x 1.25 m x 1.5 m = 1.13 m^3

3.3 Calculation of actual volume of septic tank, m^3 – Column (9)
Actual volume of septic tank = (inner length chosen, m + length of second chamber chosen, m) x (minimum water depth at outlet point chosen, m) x (inner width of septic tanks chosen, m)
Actual volume of septic tank = (1.5 m + 1 m) x 1.25 m x 1.25 m = 3.9 m^3

3.4 Calculation of biogas production, m^3/d –Column (10)
A ratio of COD to methane equal to 0.35 L/g COD at standard conditions was used in the equation proposed by Sasse (1998) to calculate the biogas production.

Biogas production = (COD_{inflow} - $COD_{outflow}$, mg/L) x daily flow, m^3/d x 0.35 L/1000 mg x 70% CH_4 x (100 – 25% dissolved CH_4)
Biogas production = (660 mg/L – 421 mg/L) x 0.3 m^3/d x 0.35 L/1000 mg x 0.7 x (1 – 0.25) = 0.03 m^3/d

4. Calculation of methane production (g/capita·d) using density of methane equal to 0.67 g/L (density at 20°C)
Methane production = biogas production, m^3/d x 10^3 L/m^3 / 1 person
Methane production = 0.03 m^3/d x 10^3 L/m^3 / 1 person = 26.9 L/capita·d x 0.67 g/L =
Methane production = 18 g/capita·d

APPENDIX E

VIAL VACUUMING PROCEDURES

(see Figure E-1)

1. Turn on the pump and then the pressure gauge.
2. Let the pressure stabilize at 12 millitorr (approximate $2.32x10^{-4}$ psi).
3. Take off the caps on the needles and close all the vials.
4. Insert one vial into each needle.
5. Open the extraction valves and wait approximately 30 sec until the pressure is again at 12 millitorr.
6. Remove the vials from the needles.
7. Repeat steps 4 - 6 for all the vials needed in the sampling event.
8. Turn off the pressure gauge and then the pump.
9. Put the caps back on the needles.

Figure E-1. View of Apparatus Used to Evacuate Vials.

APPENDIX F

SAMPLING FORMS FOR THE DIFFERENT COMPONENTS OF THE SEPTIC SYSTEM

SAMPLING FROM LIQUID SURFACE

UC DAVIS AGGIES

Date: ______________ Hour: ______________

SITE NUMBER ______________

Sample Volume ____24____ ml Time interval ____2____ min

CHARACTERISTIC	INLET	OUTLET
Scum		

Water temperature 0ft______ 1ft______ 2ft______ 3ft______ 4ft______

Water temperature 0ft______ 1ft______ 2ft______ 3ft______ 4ft______

Inlet (in) outlet (out) Duplicate (D)	Sample (number)	Time	Start time (min:sec)	Final time (min:sec)	Temperature °C
		0			
		1			
		2			
		3			
		4			
		0			
		1			
		2			
		3			
		4			

Water Quality	pH	Dissolved solids (mg/L)	Redox (mV)
Inlet			
Outlet			

Comments:

SAMPLING FROM SOIL SURFACE

Date: ______________ Hour: ______________

SITE NUMBER ______________

Sample Volume 24 ml Time interval 10 min

Insert number	Sample (number)	Time	Start time (min:sec)	Final time (min:sec)	Temperature °C
1		0			
		1			
		2			
		3			
2		0			
		1			
		2			
		3			
3		0			
		1			
		2			
		3			
4		0			
		1			
		2			
		3			
5		0			
		1			
		2			
		3			
6		0			
		1			
		2			
		3			

Comments:

SAMPLING FROM VENT SYSTEM

UCDAVIS AGGIES

Date: ________________ Hour: ________________

SITE NUMBER ________________

Sample Volume 24 ml Time interval 2 min

Vent number	Sample (number)	Start time (min:sec)	Final time (min:sec)	Air Velocity (m/s)	Temperature °C
1					
2					

Sample (number)	Pressure (Hpa)	Wind Velocity (m/s)	Outdoor Temperature °C
1			
2			
3			
4			
5			
6			

Comments:

APPENDIX G

SAMPLE OF LINEAR FIT FOR METHANE AND CARBON DIOXIDE FLUXES

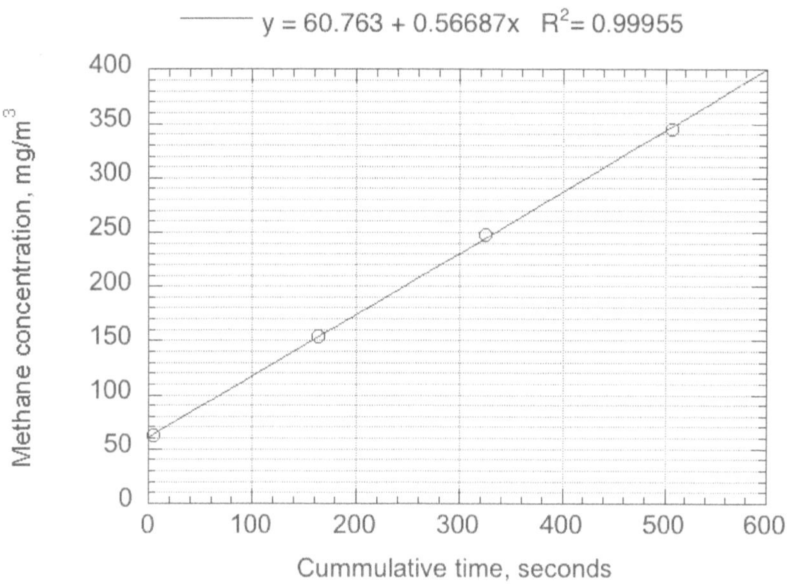

(a)

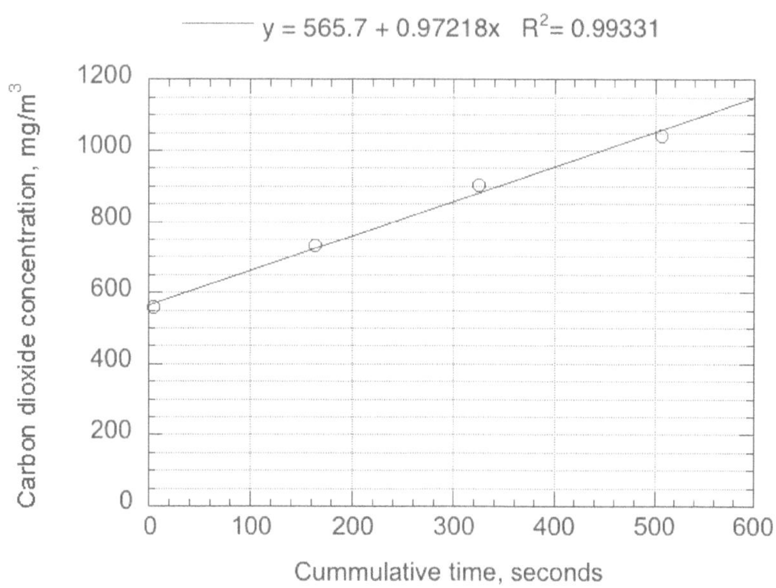

(b)

Figure G-1. Sample of Linear Fit for Methane and Carbon Dioxide Fluxes:(a) Methane and (b) Carbon Dioxide Concentration in mg/m³ Linear Fit.

APPENDIX H

FIELD DATA

H-1 Site 1

This site was characterized for a thin patchy scum layer, less than one inch. Invertebrates were not present on the scum layer. The sludge depth in the first and second compartment was 15 and 8", respectively. The sludge in the first compartment was compacted, very black, while in the second compartment was light and brownish. It was very common to observed turnover episodes. A sampling event was performed after one of the turnover events to determine the influence of these episodes in the gas emissions. The summary of the average GHG emission rates from the septic tank calculated in this site are presented in Table H-1.

Table H-1. Average of GHG Emission Rates From the Septic Tank Liquid Surface.

Date	Sample location	Gas measurement (g/capita·d)			Average liquid Temperature (°C)
		CH_4	CO_2	N_2O	
9/24/09	ST1-1-A[a]	12.00	37.85	0.0	21
	ST1-2-S[b]	6.75	15.25	0.0	22
	ST1-TE[c]	18.75	53.10	0.0	
10/07/09	ST1-1-A	20.71	55.33	0.0	21
	ST1-2-S	8.91	12.68	0.01	21
	ST1-TE	29.62	68.01	0.01	
11/05/09	ST1-1-A	9.17	67.54	0.03	18
	ST1-1-S[d]	27.54	46.09	0.0	18
	ST1-2-S	3.44	14.17	0.0	17
	ST1-TE	12.61	81.71	0.03	
	ST1-TE-1S[e]	30.98	60.26	0.0	
11/10/09	ST1-1-A	32.47	56.33	0.01	17
	ST1-2-S	3.89	17.88	0.0	20
	ST1-TE	36.36	74.21	0.01	

Date	Sample location	Gas measurement (g/capita·d)			Average liquid Temperature (°C)
		CH_4	CO_2	N_2O	
11/12/09	ST1-1-A	14.61	21.87	0.02	16
	ST1-1-S	5.30	25.14	0.0	16
	ST1-2-S	3.11	13.05	0.0	16
	ST1-TE	17.72	34.92	0.02	
	ST1-TE-1S	8.41	38.19	0.0	
11/19/09	ST1-1-A	43.09	52.19	0.01	15
	ST1-1-A-E[f]	12.84	53.89	0.0	15
	ST1-2-S	5.48	10.43	0.0	15
	ST1-TE	48.56	62.62	0.01	
	ST1-TE-1E[g]	18.32	64.32	0.0	
12/03/09	ST1-1-A	7.74	44.60	0.0	14
	ST1-1-S	3.50	15.89	0.0	14
	ST1-2-S	2.21	11.76	0.0	14
	ST1-TE	9.94	56.36	0.0	
	ST1-TE-1S	5.71	27.64	0.0	
Mean value[h]		17.87	57.80	0.00	
Standard deviation[h]		1.94	1.29	5.06	

[a] Septic tank first compartment using insert.
[b] Septic tank second compartment surface. The sample was always taken directly to the liquid surface.
[c] Septic tank total of gas emission rates from both compartments.
[d] Septic tank first compartment surface. The sample was taken directly to the liquid surface.
[e] Septic tank total of gas emission rates using values taken directly to the liquid surface.
[f] Septic tank first compartment using insert after a turnover event.
[g] Septic tank total gas emission rates using the value measured after a turnover event.
[h] Geometric mean and standard deviation values were calculated using the total emission rate value in each date.

Additionally to the gas measurements, several water quality parameters were also measured in both inlet and outlet chambers of the septic tank. The parameters measured included COD, dissolved solids, pH and redox potential. The results of the measurements obtained are summarized in Table H-2.

Table H-2. Summary of the Water Quality Results From Site 1.

Date	Sample location	Parameter			
		COD (mg/L)	pH	Redox potential (mV)	Dissolved solids (mg/L)
11/05/09	ST1-1-M[a]	-	6.69	- 150	465
	ST1-2-M[b]	-	-	-	-
11/12/09	ST1-1-M	-	6.69	- 210	455
	ST1-2-M	-	6.79	- 211	463
11/19/09	ST1-1-M	225	6.80	-170	449
	ST1-2-M	170	6.85	-180	450
12/03/09	ST1-1-M	289	6.83	- 185	473
	ST1-2-M	241	6.82	-206	480
Mean value[c]		257	6.75	-188	459
Standard deviation[c]		45	0.07	20	12
Mean value[d]		206	6.82	-199	464
Standard deviation[d]		-	0.03	17	15

[a] Septic tank first compartment. The samples were taken one foot below the liquid surface.
[b] Septic tanks second compartment. The samples were taken one foot below the liquid surface.
[c] Mean value and standard deviation from the values in the first compartment.
[d] Mean value and standard deviation from the values in the second compartment

Water samples were also collected at both compartments of the septic tank to determine the amount of methane, carbon dioxide and nitrous oxide present in solution. A summary each of the dissolved gases calculated per compartment are presented in Table H-3.

Table H-3. Summary of Dissolved Methane Measured in Each Compartment of the Septic Tank.

Date	Sample	Gas measurement (g/capita·d)		
		Methane	Carbon dioxide	Nitrous oxide
11/12/09	ST1-1-L[a]	2.71	5.63	0.0
	ST1-2-L[b]	1.12	4.55	0.0
11/19/09	ST1-1-L	1.44	5.21	0.02
	ST1-2-L	1.14	6.74	0.02
12/03/09	ST1-1-L	1.65	7.61	0.02
	ST1-2-L	0.06	0.50	0.02
Mean value[a]		1.93	6.15	0.01
Mean value[b]		0.77	3.93	0.01

[a] Septic tank first compartment.
[b] Septic tank second compartment.

After the initial inspections, Site 1 was selected for venting and soil dispersal system sampling. Samples from the vent system were taken at two different cleanout vents located before and after the septic tank. An anemometer was used to determine the air flow inside the vent pipe to calculate the flow rate and therefore the GHG emission rates. The air velocity in the venting pipes ranged from 35 to 100 ft/min. A summary of the average GHG emission rates from the vent system are presented in Table H-4.

Table H-4. Average of GHG Emission Rates From the Vent System.

Date	Sample location	Gas measurement (g/capita·d)			Average Temperature (°C)
		CH_4	CO_2	N_2O	
11/10/09	V-1-1[a]	12.11	710.65	0.49	17
11/12/09	V-1-1	8.43	744.30	0.95	15
	V-1-2[b]	0.03	637.43	0.52	16
11/19/09	V-1-1	9.72	393.03	0.42	15
	V-1-2	0.03	10.93	0.02	14
12/03/09	V-1-1	9.17	533.10	0.42	13
	V-1-2	0.0	69.21	0.09	14
02/05/10	V-1-1	6.49	446.01	0.18	11
	V-1-2	0.06	432.79	0.18	10
Mean value[c]		8.44	527.00	0.37	
Standard deviation[c]		0.01	103.62	0.08	
Mean value[d]		1.27	1.31	1.82	
Standard deviation[d]		6.60	0.97	6.09	

[a] Cleanout vent located before the septic tank.
[b] Cleanout vent located before the soil dispersal system
[c] Geometric mean value and standard deviation calculated to the cleanout vent located before the septic tank
[d] Geometric mean value and standard deviation calculated to the cleanout vent located before the soil dispersal system.

The main soil dispersal system characteristics are summarized in Table H-5. Six 12" inserts were installed to sample from the soil dispersal system. Five were distributed above the system and one was inserted outside the system for control purposes.

Table H-5. General Characteristics of the Soil Dispersal System.

Characteristic	Value	Unit
Area	205	m^2
Slope	20	%
Number of trenches	3	-
Rings installed	6[a]	-

[a] Five distributed on the trenches and one located 10 ft away from the dispersal system area for control purposes.

Gas samples were taken using the flux chamber and were analyzed for methane, carbon dioxide and nitrous oxide. The system was sampled twice to verify the results obtained in the first set of measurements. On the second trial, emission rates of the three GHGs were detected in the control ring and also in two of rings located above the dispersal system. However, the values calculated were similar or less than the control value as seen in Table H-6.

Table H-6. GHG Emission Rates From the Soil Dispersal System.

Date[a]	Sample location	Gas measurement (g/capita·d) CH_4	CO_2	N_2O	Average Temperature (°C)
12/03/09	SD1-R1C[b]	0.004	485	0.04	12
	SD1-R2[c]	ND[d]	ND	ND	11
	SD1-R3	ND	ND	0.00	11
	SD1-R4	ND	236	ND	9
	SD1-R5	ND	ND	ND	10
	SD1-R6	ND	671	0.1	10

[a] On 11/12/09 a sampling event was performed at this site. However, fluxes of CH_4, CO_2 and N_2O were not detected in the 5 sampling rings. Control values for CH_4, CO_2 and N_2O in g/capita·d were 0.8, 0 and 0 respectively.
[b] Control ring located on the soil dispersal system.
[c] Sampling ring located on the soil dispersal system.
[d] Flux of gas was not detected.

H-2 Site 2

The scum layer in the first compartment of this septic tank was thick and compacted approximately 4".

Table H-7. Average of GHG Emission Rates From the Septic Tank Liquid Surface.

Date	Sample location	Gas measurement (g/capita·d)			Average liquid Temperature (°C)
		CH_4	CO_2	N_2O	
9/24/09	ST2-1-A[a]	1.04	39.31	0.03	26
	ST2-2-S[b]	3.07	14.18	0.0	26
	ST2-TE[c]	4.11	53.50	0.03	
10/07/09	ST2-1-A	12.34	70.51	0.0	25
	ST2-2-S	0.51	12.71	0.0	24
	ST2-TE	12.85	83.23	0.0	
11/05/09	ST2-1-A	10.77	58.86	0.0	22
	ST2-2-S	1.11	6.53	0.0	20
	ST2-TE	11.88	65.39	0.0	
11/19/09	ST2-1-A	1.56	36.24	0.02	18
	ST2-2-S	2.11	12.52	0.01	18
	ST2-TE	3.67	48.75	0.02	
12/08/09	ST2-1-A	4.81	19.90	0.0	14
	ST2-2-S	17.06	29.16	0.0	14
	ST2-TE	21.87	49.06	0.0	
Mean value[d]		8.72	58.69	0.01	
Standard deviation[d]		2.17	1.26	3.25	

[a] Septic tank first compartment using insert.
[b] Septic tank second compartment using insert.
[c] Septic tank total gas emission rates from both compartments.
[d] Geometric mean value and standard deviation calculated using the total of gas emission rates from both compartments.

It was black with humus like appearance, with larvae and small flies present on the top. The sludge depth in the first and second compartment was 12 and 6", respectively. The summary of the GHG emission rates from the septic tank in this site is presented in Table H-7

Water quality parameters were also measured in both inlet and outlet chambers of the septic tank. The parameters measured included COD, dissolved solids, pH and redox potential. The results of the measurements obtained are summarized in Table H-8.

Table H-8. Summary of the Water Quality Results from Site 2.

Date	Sample location	COD (mg/L)	pH	Redox potential (mV)	Dissolved solids (mg/L)
		Parameter			
11/05/09	ST2-1-M[a]	-	7.24	-241	636
	ST2-2-M[b]	-	7.20	-222	652
11/19/09	ST2-1-M	254	6.87	-195	620
	ST2-2-M	173	7.06	-230	612
12/08/09	ST2-1-M	240	7.02	-190	612
	ST2-2-M	175	7.17	-220	632
Mean value[c]		247	7.04	-209	623
Standard deviation[c]		-	0.19	28.11	12.22
Mean value[d]		174	7.14	-224	632
Standar deviation[d]		-	0.07	5.29	20

[a] Septic tank first compartment. The samples were taken one foot below the liquid surface.
[b] Septic tanks second compartment. The samples were taken one foot below the liquid surface.
[c] Mean value and standard deviation from the values in the first compartment.
[d] Mean value and standard deviation from the values in the second compartment.

Water samples were also collected at both compartments of the septic tank to determine the amount of methane, carbon dioxide and nitrous oxide present in solution. A summary of the dissolved gases calculated per compartment are presented in Table H-9.

Table H-9. Summary of Dissolved Methane Measured in Each Compartment of the Septic Tank.

Date	Sample	Methane	Carbon dioxide	Nitrous oxide
		Gas measurement (g/capita·d)		
11/19/09	ST2-1-L[a]	2.05	7.51	0.02
	ST2-2-L[b]	1.42	4.90	0.02
12/08/09	ST2-1-L	1.59	4.46	0.0
	ST2-2-L	0.004	0.29	0.0
Mean value[a]		1.82	6.23	0.01
Mean value[b]		0.71	2.60	0.01

[a]Septic tank first compartment.
[b] Septic tank second compartment

Site 2 was also selected to be sampled in detail. However, the soil infiltration system was not a good candidate for gas sampling due to the dense vegetation above it. Therefore, gas samples were just taken at the septic tank and vent system. Samples from the vent system were taken at a cleanout vent located before the septic tank. The air velocity of the air in the venting pipes varied from 30 to 80 ft.min. A summary of the GHG emission rates from the vent system is presented in Table H-10.

Table H-10. GHG Emission Rates From the Vent System[a].

Date	Sample location	Gas measurement (g/capita·d)			Average Temperature (°C)
		CH_4	CO_2	N_2O	
9/24/09	V-2-1	18.58	249.03	0.16	19

[a]The vent system was sampled twice. However, during the second trial on December 8th, the vent pipe was frozen (ambient temperature was 6 °C) and the velocity inside the pipe was zero.

H-3 Site 3

In this site gas samples were only taken at the septic tank liquid surface. The scum layer in this site was thick between 3 and 4", but it was not as compacted as the scum layer at Sites 1 and 2. Contrary, this scum layer has a crumbly earth appearance and it was populated by earthworms. The sludge depth in the first and second compartment was 12 and 15", respectively. The summary of the average GHG emission rates from the septic tank calculated in this site is presented in Table H-11.

Table H-11. Average of GHG Emission Rates From the Septic Tank Liquid Surface.

Date	Sample location	Gas measurement (g/capita·d)			Average liquid Temperature (°C)
		CH_4	CO_2	N_2O	
9/24/09	ST3-1-A[a]	0.87	26.06	0.04	27
	ST3-2-A[b]	1.55	20.47	0.0	26
	ST3-TE[c]	2.43	46.53	0.04	
10/07/09	ST3-1-A	24.06	54.66	0.0	23
	ST3-2-A	1.03	7.77	0.01	22
	ST3-TE	25.09	62.43	0.01	
11/05/09	ST3-1-A	12.39	33.39	0.06	19
	ST3-2-A	1.31	9.75	0.01	19
	ST3-TE	13.70	43.14	0.07	
Mean value[d]		9.42	50.04	0.03	
Standard deviation[d]		3.36	1.22	2.38	

[a] Septic tank first compartment using insert.
[b] Septic tank second compartment using insert.
[c] Septic tank total gas emission rates from both compartments.
[d] Geometric mean value and standard deviation calculated using the total gas emission rates from both compartments.

Water quality parameters were also measured in both inlet and outlet chambers of the septic tank. The results of the measurements obtained are summarized in Table H-12.

Table H-12. Summary of the Water Quality Results from Site 3.

Date	Sample location	pH	Redox potential (mV)	Dissolved solids (mg/L)
		Parameter		
11/05/09	ST3-1-M[a]	6.82	-215	499
	ST3-2-M[b]	7.20	-272	755

[a] Septic tank first compartment. The sample was taken one foot below the liquid surface.
[b] Septic tanks second compartment. The sample was taken one foot below the liquid surface.

H-4 Site 4

As soon as the tank lids were opened, a pine odor was detected. The sludge depth in the tank was 14", it had a brownish color and it was very light. The summary of the average GHG emission rates from the septic tank calculated in this site are presented in Table H-13.

Table H-13. Average of GHG Emission Rates from the Septic Tank Liquid Surface.

Date	Sample location	Gas measurement (g/capita·d)			Average liquid Temperature (°C)
		CH_4	CO_2	N_2O	
9/24/09	ST4-1-S[a]	0.55	2.70	0.0	22
	ST4-2-S[b]	0.09	3.39	0.0	22
	ST4-TE[c]	0.63	6.09	0.0	
10/07/09	ST4-1-S	0.17	3.42	0.0	18
	ST4-2-S	0.01	1.87	0.0	19
	ST4-TE	0.17	5.30	0.0	
11/05/09	ST4-1-S	0.06	0.21	0.0	15
	ST4-2-S	0.01	0.89	0.0	15
	ST4-TE	0.07	1.09	0.0	
Mean value[d]		0.20	3.28	0.0	
Standard deviation[d]		3.03	2.60	0.0	

[a] Septic tank first compartment. It was not necessary to install an insert to sample from the first compartment.
[b] Septic tank second compartment. It was not necessary to install an insert to sample from the second compartment.
[c] Septic tank total gas emission rates from both compartments.
[d] Geometric mean value and standard deviation calculated using the total gas emission rates from both compartments.

Water quality parameters were also measured in both inlet and outlet chambers of the septic tank. Measurements were also made to the water supply of the house. The results of the measurements obtained are summarized in Table H-14.

Table H-14. Summary of the Water Quality Results from Site 4.

Date	Sample location	Parameter		
		pH	Redox potential (mV)	Dissolved solids (mg/L)
11/05/09	ST4-1-M[a]	6.70	+131	46
	ST4-2-M[b]	6.26	+216	79
	WSH[c]	8.3	+530	29

[a] Septic tank first compartment. The sample was taken one foot below the liquid surface.
[b] Septic tanks second compartment. The sample was taken one foot below the liquid surface.
[c] Sample taken to the water supply of the house.

H-5 Site 5

The scum layer was black, very thin, around 1 inch, and populated with larvae. Floating toilet paper was very common in this septic tank. The sludge was brownish and very light with a depth of 8 and 6" in the first and second compartment, respectively. The average GHG emission rates from the septic tank calculated in this site are presented in Table H-15.

Table H-15. Average of GHG Emission Rates from the Septic Tank Liquid Surface.

Date	Sample location	Gas measurement (g/capita·d)			Average liquid Temperature (°C)
		CH_4	CO_2	N_2O	
9/24/09	ST5-1-A[a]	3.01	13.65	0.0	25
	ST5-2-S[b]	1.23	5.79	0.0	24
	ST5-TE[c]	4.25	19.44	0.0	
10/07/09	ST5-1-A	4.94	27.02	0.0	21
	ST5-2-S	0.66	3.86	0.0	21
	ST5-TE	5.59	30.88	0.0	
11/05/09	ST5-1-A	9.88	39.66	0.0	17
	ST5-2-S	0.64	4.23	0.02	16
	ST5-TE	10.52	43.89	0.02	
Mean value[d]		6.30	29.76	0.01	
Standard deviation[d]		1.59	1.40	1.59	

[a] Septic tank first compartment using insert.
[b] Septic tank second compartment. It was not necessary to install an insert to sample in the second compartment.
[c] Septic tank total gas emission rates from both compartments.
[d] Geometric mean value and standard deviation calculated using the total gas emission rates from both compartments.

Water quality parameters were also measured in both inlet and outlet chambers of the septic tank. The results of the measurements obtained are summarized in Table H-16.

Table H-16. Summary of the Water Quality Results from Site 5.

Date	Sample location	Parameter: pH	Redox potential (mV)	Dissolved solids (mg/L)
11/05/09	ST5-1-M[a]	7.05	-205	690
	ST5-2-M[b]	7.42	-223	749

[a] Septic tank first compartment. The sample was taken one foot below the liquid surface.
[b] Septic tanks second compartment. The sample was taken one foot below the liquid surface.

H-6 Site 6

A thick scum layer of around 3" was observed in the liquid close the inlet tee; however the rest of the tank had a patchy brownish scum layer. The outlet water surface had a thin oily layer. The sludge depth measured in the first and second compartment was 6 and 20", respectively. The summary of the average GHG emission rates from the septic tank calculated in this site are presented in Table H-17.

Table H-17. Average of GHG Emission Rates from the SepticTank Lquid Surface.

Date	Sample location	Gas measurement (g/capita·d)			Average liquid Temperature (°C)
		CH_4	CO_2	N_2O	
9/24/09	ST6-1-A[a]	8.31	17.37	0.01	22
	ST6-2-A[b]	1.64	13.39	0.0	22
	ST6-TE[c]	9.95	30.76	0.01	
11/05/09	ST6-1-A	5.79	60.02	0.0	16
	ST6-2-A	1.64	11.40	0.0	15
	ST6-TE	7.43	71.42	0.0	
Mean value[d]		8.60	46.87	0.0	
Standard deviation[e]		1.23	1.82	1.48	

[a] Septic tank first compartment using insert.
[b] Septic tank second compartment using insert.
[c] Septic tank total gas emission rates from both compartments.
[d] Geometric mean value calculated using the total gas emission rates from both compartments.
[e] Geometric standard deviation calculated using the total gas emission rates from both compartments.

Water quality parameters were also measured in both inlet and outlet chambers of the septic tank. The results of the measurements obtained are summarized in Table H-18.

Table H-18. Summary of the Water Quality Results from Site 6.

Date	Sample location	pH	Redox potential (mV)	Dissolved solids (mg/L)
		Parameter		
11/05/09	ST6-1-M[a]	6.93	-212	673
	ST6-2-M[b]	6.98	-197	691

[a] Septic tank first compartment. The sample was taken one foot below the liquid surface.
[b] Septic tanks second compartment. The sample was taken one foot below the liquid surface

H-7 Site 7

The scum layer had an approximate thickness of 3", and it tend to accumulates in the around the inlet tee. The scum was black, had a crumbly earth appearance in some spots and it was populated by larvae and small flies. The sludge depth in the first and second compartment was 14 and 6 ", respectively. The summary of the average GHG emission rates from the septic tank calculated in this site are presented in Table H-19.

Water quality parameters were also measured in both inlet and outlet chambers of the septic tank. The parameters measured included COD, dissolved solids, pH and redox potential. The results of the measurements obtained are summarized in Table H-20.

Table H-19. Average of GHG Emission Rates from the Septic Tank Liquid Surface.

Date	Sample location	Gas measurement (g/capita·d) CH_4	CO_2	N_2O	Average liquid Temperature (°C)
10/07/09	ST7-1-A[a]	5.35	23.75	0.0	21
	ST7-2-A[b]	0.89	6.0	0.0	20
	ST7-TE[c]	6.24	29.75	0.0	
11/05/09	ST7-1-A	24.23	36.01	0.0	16
	ST7-2-A	0.79	2.92	0.0	17
	ST7-TE	25.01	38.93	0.0	
11/17/09	ST7-1-A	19.20	43.09	0.0	17
	ST7-2-A	0.58	7.19	0.0	16
	ST7-TE	19.78	50.28	0.0	
12/01/09	ST7-1-A	26.25	50.56	0.02	15
	ST7-2-A	0.58	5.52	0.0	14
	ST7-TE	26.83	56.07	0.02	
Mean value[d]		16.97	42.51	0.0	
Standard deviation[d]		1.86	1.33	-	

[a] Septic tank first compartment using insert.
[b] Septic tank second compartment using insert.
[c] Septic tank total gas emission rates from both compartments.
[d] Geometric mean value and standard deviation calculated using the total gas emission rates from both compartments.

Table H-20. Summary of the Water Quality Results from Site 7.

Date	Sample location	COD (mg/L)	pH	Redox potential (mV)	Dissolved solids (mg/L)
11/05/09	ST7-1-M[a]	-	6.61	-201	649
	ST7-2-M[b]	-	6.89	-180	682
11/17/09	ST7-1-M	-	6.52	-170	560
	ST7-2-M	-	6.74	-220	590
12/01/09	ST7-1-M	408	6.56	-196	607
	ST7-2-M	306	6.69	-195	604
Mean value[c]		-	6.56	-189	605
Standard deviation[c]		-	0.05	17	45
Mean value[d]		-	6.77	-198	625
Standard deviation[d]		-	0.10	20	50

[a] Septic tank first compartment. The sample was taken one foot below the liquid surface.
[b] Septic tanks second compartment. The sample was taken one foot below the liquid surface.
[c] Mean value and standard deviation from the values in the first compartment.
[d] Mean value and standard deviation from the values in the second compartment.

Water samples were also collected at both compartments of the septic tank to determine the amount of methane, carbon dioxide and nitrous oxide present in solution. A summary of the dissolved gases calculated per compartment are presented in Table H-21.

Table H-21. Summary of Dissolved Methane Measured in Each Compartment of the Septic Tank.

Date	Sample	Gas measurement (g/capita·d) Methane	Carbon dioxide	Nitrous oxide
11/17/09	ST7-1-L[a]	0.29	1.88	0.01
	ST7-2-L[b]	0.0	0.17	0.01
12/01/09	ST7-1-L	0.94	6.6	0.0
	ST7-2-L	0.0	0.18	0.0
Mean value[a]		0.61	4.24	0.005
Mean value[b]		0.0	0.17	0.0

[a] Septic tank first compartment.
[b] Septic tank second compartment

Site 7 was selected to be sampled in more detailed. Therefore, gas samples were taken at the septic tank, vent and soil dispersal systems. Samples from the vent system were taken at the cleanout vent located before the septic tank. Air velocity measured in the passive house venting systems ranged from 40 to 100 ft/min when an effluent filter was not present. Effluent filters reduced the air flow through the vent system to a range of 0 to 0.1 m/s (0 to 10 ft/min). The results from the vent systems measurements are summarized in Table H-22.

Table H-22. GHG Emission Rates from the Vent System.

Date	Sample location	Gas measurement (g/capita·d)			Average Temperature (°C)
		CH_4	CO_2	N_2O	
11/17/09	V7-1EF[a]	31.73	192.56	0.21	14
12/01/09	V7-1EF	7.53	57.73	0.01	17
	V7-1OF[b]	5.51	36.18	0.0	13
	V7-1WEF[c]	14.89	115.75	0.02	12
Mean value[a]		13.39	93.47	0.04	

[a] The vent system samples were typically taken without opening the tank lids and with the installed effluent filter. The mean was calculated using the results from the two sampling events performed in this conditions.
[b] Vent sample taken using a biotube effluent filter model FTi0418-S from Orenco Systems, Inc.
[c] Vent sample taken without the effluent filter and tank lids off.

The main soil dispersal system characteristics are summarized in Table H-23. Six 12" inserts were installed to sample from the soil dispersal system. Five were distributed above the system and one was inserted outside the system for control purposes.

Table H-23. General Characteristics of the Soil Dispersal System.

Characteristic	Value	Unit
Area	650	m^2
Slope	8.5	%
Number of trenches	3	-
Rings installed	6[a]	-

[a] Five distributed on the trenches and one located 6 ft away from the dispersal system area for control purposes.

Gas samples were taken and analyzed for methane, carbon dioxide and nitrous oxide. During the first set of measurements, methane gas was detected once in one of the rings but not in the control one. In the second sampling date, the carbon dioxide emission rates values were similar or less than the control value and nitrous oxide emission rates were double than the control value but the amount is not significant. A summary of the results from the soil dispersal system at Site 7 are found in Table H-24.

Table H-24. GHG Emission Rates from the Soil Dispersal System.

Date	Sample location	Gas measurement (g/capita·d) CH$_4$	CO$_2$	N$_2$O	Average Temperature (°C)
11/17/09	SD7-R1[a]	ND[c]	ND	ND	16.8
	SD7-R2C[b]	ND	619	0.19	14.8
	SD7-R3	ND	ND	ND	17.2
	SD7-R4	ND	110	0.11	19.4
	SD7-R5	ND	ND	ND	19.1
	SD7-R6	0.0	ND	0.10	16.5
12/01/09	SD7-R1	ND	842	0.07	16.8
	SD7-R2C	ND	843	0.04	14.8
	SD7-R3	0.09	469	0.11	17.2
	SD7-R4	ND	ND	ND	19.4
	SD7-R5	ND	ND	0.5	19.1
	SD7-R6	ND	851	0.16	16.5

[a] Sampling ring located on the soil dispersal system.
[b] Control ring located on the soil dispersal system.
[c] Flux of gas was not detected.

H-8 Site 8

The scum layer in this septic tank was thin, less than one inch, it was black and small larvae were observed in both compartments. The sludge depth both compartments of the septic tank was 3' and it had a very thick consistency. The total water depth was 5'. therefore the clear zone of the tank was reduced to 2'. The tank has a strong odor compared to the tanks in ALT. A summary of the GHG emission rates from the septic tank calculated in this site are shown in Table H-25.

Table H-25. Average of GHG Emission Rates from the Septic Tank Liquid Surface.

Date	Hour	Sample location	Gas measurement (g/capita·d) CH$_4$	CO$_2$	N$_2$O	Average liquid Temperature (°C)
10/21/09	11:00 am	ST8-1-A[a]	9.88	46.58	0.0	17
		ST8-2-A[b]	1.10	8.31	0.0	17
		ST8-TE[c]	10.98	54.89	0.0	
	3:00pm	ST8-1-A	10.96	25.59	0.0	17
		ST8-2-A	1.14	6.36	0.0	17
		ST-8-TE	12.10	31.94	0.0	
	7:00pm	ST8-1-A	1.14	13.51	0.0	16
		ST8-2-A	2.80	1.97	0.0	16
		ST8-TE	3.44	15.48	0.0	
	11:00pm	ST8-1-A	73.72	33.54	0.0	16
		ST8-2-A	1.95	12.57	0.0	16
		ST8-TE	75.68	46.21	0.0	
10/22/09	3:00 am	ST8-1-A	10.06	74.54	0.0	16
		ST8-2-A	1.21	1.00	0.0	16

Date	Hour	Sample location	Gas measurement (g/capita·d)			Average liquid Temperature (°C)
			CH_4	CO_2	N_2O	
		ST8-TE	11.27	75.54	0.0	
	7:00 am	ST8-1-A	34.41	10.92	0.0	16
		ST8-2-A	1.48	17.45	0.0	16
		ST8-TE	35.89	28.37	0.0	
	11:00am	ST8-1-A	ND[d]	ND	ND	17
		ST8-2-A	4.05	5.52	0.02	17
		ST8-TE	4.05	5.52	0.02	
12/10/09	7:30 am	ST8-1-A	1.33	12.35	0.01	13
		ST8-2-A	1.38	2.43	0.0	13
		ST8-TE	2.72	14.77	0.01	
	11:00 am	ST8-1-A	2.70	1.62	0.01	13
		ST8-2-A	1.42	2.17	0.0	13
		ST8-TE	4.11	3.79	0.01	
	2:00pm	ST8-1-A	10.24	1.80	0.0	13
		ST8-2-A	0.98	0.72	0.01	13
		ST8-TE	11.23	2.53	0.01	
	7:00pm	ST8-1-A	0.34	0.54	0.0	13
		ST8-2-A	0.68	0.73	0.01	13
		ST8-TE	1.02	1.26	0.01	
12/11/09	8:00am	ST8-1-A	4.63	1.74	0.0	12
		ST8-2-A	1.03	0.86	0.0	12
		ST8-TE	5.66	2.60	0.0	
Mean value[e]			8.99	12.58	0.01	
Standard deviation[e]			3.33	3.9	4.42	

[a] Septic tank first compartment using insert.
[b] Septic tank second compartment. The flux chamber was installed directly to the liquid surface.
[c] Septic tank total gas emission rates from both compartments.
[d] Not detected.
[e] Geometric mean value and standard deviation calculated using the total gas emission rates.

Water quality parameters were also measured in both inlet and outlet chambers of the septic tank. The parameters measured included COD, dissolved solids, pH and redox potential. The results of the measurements obtained are summarized in Table H-26.

Table H-26. Summary of the Water Quality Results from Site 5[a].

		Parameter			
Date	Sample location	COD (mg/L)	pH	Redox potential (mV)	Dissolved solids (mg/L)
12/10/09	ST7-1-M[a]	164	7.66	-	1113
	ST7-2-M[b]	108	7.57	-212	1200
12/11/09	ST7-1-M	168	-	-	-
	ST7-2-M	112	-	-	-

[a] Septic tank first compartment. The sample was taken one foot below the liquid surface.
[b] Septic tanks second compartment. The sample was taken one foot below the liquid surface.

Water samples were also collected at both compartments of the septic tank to determine the amount of methane present in solution. A summary of the dissolved methane calculated per compartment are shown in Table H-27.

Table H-27. Summary of Dissolved Methane Measured in Each Compartment of the Septic Tank.

		Gas measurement (g/capita·d)		
Date	Sample	Methane	Carbon dioxide	Nitrous oxide
12/10/09	ST8-1-L[a]	0.50	2.14	0.002
	ST8-2-L[b]	0.33	1.61	0.001
12/11/09	ST8-1-L	0.55	2.72	0.002
	ST8-2-L	0.42	2.39	0.001
Mean value[a]		0.52	2.43	0.002
Mean value[b]		0.38	2.00	0.001

[a] Septic tank first compartment.
[b] Septic tank second compartment.

H-9 Summary of Results

A summary of the sampling dates, calculated average GHG emission rates value and the standard deviation from the different septic tanks studied is presented in Table H-28.

Table H-28. Summary of GHG Emission Rates from the Eight Septic Tanks Used in the Study.

Septic tank number	Gas Measurement (g/capita·d)					
	CH_4		CO_2		N_2O	
	TGE[a]	SD[b]	TGE	SD	TGE	SD
1	17.87	1.94	57.80	1.29	0.0	5.06
2	8.72	2.17	58.69	1.26	0.01	3.25
3	9.42	3.36	50.04	1.22	0.03	2.38
4	0.20	3.03	3.28	2.60	0.0	0.0
5	6.30	1.59	29.76	1.40	0.01	1.59
6	8.60	1.23	46.87	1.82	0.0	1.48
7	16.97	1.86	42.51	1.33	0.0	-
8	8.99	3.33	12.58	3.90	0.01	4.42

[a] Mean value of total gas emission rates from both compartments.
[b] Standard deviation calculated to the total gas emission rates from both compartments.

APPENDIX I

CALCULATION OF COD IN TERMS OF CO_2 FOR MASS BALANCE ON THE SOIL DISPERSAL SYSTEM

1. Balanced oxidation equation for wastewater
   $2C_{10}H_{19}O_3N$ + $25\ O_2$ = $20CO_2$ + $16H_2O$ + $2NH_3$
   (221 g/mol) (32 g/mol) (44g /mol)

2. Theoretical (chemical) oxygen demand for $C_{10}H_{19}O_3N$
   (25 x 32)/(2 x 221) = 1.81 g O_2/ g $C_{10}H_{19}O_3N$ = 1.81 g COD/ g $C_{10}H_{19}O_3N$

3. Carbon dioxide equivalent for $C_{10}H_{19}O_3N$
   (20 x 44)/(2 x 221) = 1.99 g CO_2/ g $C_{10}H_{19}O_3N$

4. Carbon dioxide produced from COD
   (1.99 g CO_2 / g $C_{10}H_{19}O_3N$) / (1.81 g COD / g $C_{10}H_{19}O_3N$) = 1.1 g CO_2/g COD

5. COD value measured at the effluent equal to 206 mg/L, assuming that the flow coming to the septic tank ranges from 200 to 590 L/capita·d (Crites and Tchobanoglous, 1998; GDPUD, 2010), the COD in terms of CO_2 is calculated.

   (206 mg COD/L) / (1 g/1000 mg) x (200 L/d) x (1.1 g CO_2/g COD)
   = 45.3 g CO_2/capita·d

   (206 mg COD/L) / (1 g/1000 mg) x (590 L/d) x (1.1 g CO_2/g COD)
   = 133.7 g CO_2/capita·d

Thus, the CO_2 equivalent of the COD loading to the soil dispersal system is expected to range 45.3 to 133.7 g CO_2/capita·d.

References

Aneja, V.P., J. Blunden, C.S. Clairborn, and H.H. Rogers (2006) *"Dynamic Chamber System to Measure Gaseous Compounds Emissions and Atmospheric-Biospheric Interactions,"* I. Barnes and K.J. Rudzinski (eds.), Environmental Simulation Chambers: Application to Atmospheric Chemical Processes, 97-109, Netherlands.

Ball, H. (2009) Personal communication.

Bogner, J., R. Pipatti, S. Hashimoto, C. Diaz, K. Mareckova, L. Diaz, P. Kjeldsen, S. Monnis, A. Faaij, Q. Gao, T. Zhang, M.S. Ahmed, R.T.M. Sutamihardja, and R. Gregory (2008) "Mitigation of Global Greenhouse Gas Emissions from Waste: Conclusions and Strategies from the Intergovernmental Panel on Climate Change (IPCC). Fourth Assessment Report. Working Group III (Mitigation)," *J. Waste Management and Research.,* **26**, 11-32.

Bounds, T.R. (1997) "Design and Performance of Septic Tanks," in M.S. Bedinger, J.S. Fleming, and A.I. Johanson (eds.), *Site Characterization and Design of On-Site Septic Systems*, ASTM STP 11324, American Society for Testing and Materials.

Bitton, G. (2005) *Wastewater Microbiology*, 3rd ed., Jhon Wiley & Sons, Inc., Hoboken, New Jersey.

California Plumbing Code (2007) IAPMO. *Chapter Nine, Vents*. Ontario, CA.
Cakir, F.Y. and M.K. Stenstrom (2005) "Greenhouse Gas Production: A Comparison Between Aerobic and Anaerobic Wastewater Treatment Technology," *Water Res.,* **39**, 4197-4203.

Chandran, K. (2009) "Characterization of Nitrogen Greenhouse Gas Emissions from Wastewater Treatment BNR Operations," *Field Protocol with Quality Assurance Plan,* Water Environment Research Foundation, Alexandria, VA.

Christensen, S., P. Ambus, J.R.M. Arah, H. Clayton, B. Galle, D.W.T. Griffith, K.J. Hargreaves, K.A. Smith, M. Welling, and F.G. Wienholds (1996) *"Nitrous Oxide Emissions from an Agricultural Field: Comparison Between Measurements by Flux Chamber and Micrometerological Techniques," Atmospheric Environment*, **30**,24, 4183-4190.

Crites, R. and G. Tchobanoglous (1998) *Small and Decentralized Wastewater Management Systems,* McGraw-Hill, New York.

Dale, R. (1982) *"The Arthropod and Related Fauna of Septic Tanks in Novato, Marin County, California,"* Master thesis, Department of Biology, California State University, Hayward, CA.

D'Amato, V.A., A. Bahe, T.R. Bounds, B. Comstock, T. Konsler, S.K. Liehr, S.C. Long, K. Ratanaphuks, C.A. Rock, and K. Sherma (2008) "Factors Affecting the Performance of Primary

Treatment in Decentralized Wastewater Treatment Systems," Water Environment Research Foundation, Alexandria, VA.

DeSutter, T.M. and J.M. Ham (2005) "Emissions and Carbon Balance Estimates of a Swine Production Facility," *J. Environ. Qual.* **34**,198-206.

Dunbar, W.P. (1908) *Principles of Sewage Treatment*, Translated by H.T. Calvert. Charles Griffin and Co., Ltd., England.

Duncan, M. and N. Horan (2003) *Handbook of Water and Wastewater Microbiology,* Academic Press, ELSEVIER, San Diego, CA.

Erickson, J. and E.J. Tyler (2000) *A Model for Soil Oxygen Delivery to Wastewater Infiltration Surfaces,* NOWRA 2000 Conference Proceedings. National Onsite Wastewater Recycling Association, pp. 91-96.

Foley, J. and P. Lant (2009) *Direct Methane and Nitrous Oxide Emissions from Full-Scale Wastewater Treatment Systems*, The University of Queensland, Australia. Water Services Association of Australia (WSAA) Occasional Paper, No. 24.

GDPUD (2009) Georgetown Divide Public Utility District, Personal communication with D. Creeks.

GDPUD (2010) Georgetown Divide Public Utility District, Personal communication with B. Siren.

Höglund, C. (2001). "*Evaluation of Microbial Health Risks Associated with the Reuse of Source Separated Human Urine,*" Ph.D. thesis, Department of Biotechnology, Royal Institute of Technology, Stockholm, Sweden.

Hutchinson, G.L. and G.P. Livingston (1993) "Use of Chamber Systems to Measure Trace Gas Fluxes," in L.A. Harper, A.R. Moiser, J.M. Duxbury, and D.E. Rolston (eds.) *Agricultural Ecosystem Effects on Trace Gases and Global Climate Change*, ASA Spec. Publ. 55. ASA, Madison, WI.

IPCC (1996) Climate Change 1995: The Science of Climate Change. Intergovernmental Panel on Climate Change, J.T. Houghton, L.G. Meira Filho, B.A. Callander, N. Harris, A. Kattenberg, and K. Maskell (eds.), Cambridge University Press, Cambridge, United Kingdom.

IPCC (2007) Climate Change (2007) Synthesis Report. Contribution of Working Groups I, II and III to the Fourth Assessment Report of the Intergovernmental Panel on Climate Change [Core Writing Team, R.K. Pachauri, and A. Reisinger (eds.)]. IPCC, Geneva, Switzerland.

Isa, Z., S. Grusenmeyer, and W. Verstraete (1986) "Sulfate Reductions Relative to Methane Production in High-Rate Anaerobic Digestion: Technical Aspects," *J. Applied and Environmental Microbiology.*, **51**,3, 572-579.

Janna, W.S. (2007) *Conduction Shape Factor Method Applied to the Modeling of Oxygen Diffusion Through Soil*, Eleventh Individual and Small Community Sewage Systems Conference Proceedings, 20-24 October 2007, Warwick, Rhode Island, USA, ASAE, St. Josephs, MI.

Kinnicutt, L.P., C.E.A. Winslow, and R.W. Pratt (1910) *Sewage Disposal*, John Wiley and Sons. New York.

Knowles, R. (1993) "Methane: Processes of Production and Consumption," in L.A. Harper, A.R. Moiser, J.M. Duxbury, and D.E. Rolston (eds.) *Agricultural Ecosystem Effects on Trace Gases and Global Climate Change*, ASA Spec. Publ. 55. ASA, Madison, WI.

Leverenz, H., G. Tchobanoglous, and J.L. Darby (2009) "Clogging in Intermittently Dosed Sand Filters Used For Wastewater Treatment," *Water Research,* **43**, 3, 695-705.

Maier, R.M., I.L. Pepper, and C.P. Gerba (2009) *Environmental Microbiology,* 2nd ed., academic Press, ELSEVIER Inc., London.

Maillacheruvu, K.Y. and G.F. Parkin (1996) "Kinetics of Growth, Substrate Utilization, and Sulfide Toxicity for Propionate, Acetate and Hydrogen Utilizers in Anaerobic Systems," *Water Environmental Research,* **68**, 7, 1099-1106.

McCarty, P.L. (1964) "Anaerobic Waste Treatment Fundamental," *Public Works*, **95,** 9, 107 112,10, 123-126, 11, 91-94, 12, 95-99.

McCarty, P.L. (2001) "The Development of Anaerobic Treatment and Its Future," *Water Sci. Tech.,* **44**, 8, 149-155.

Moore, T.R. and N.T. Roulet (1991) "A Comparison of Dynamic and Static Chambers for Methane Emissions Measurements From Subartictic Fens," *Atmos. Ocean,* **29, 1,** 102-109.

Patel, G.B., A.W. Khan, and L.A. Roth (1978) "Optimum Levels of Sulfate and Iron for the Cultivations of Pure Cultures of Methanogens in Synthetic Media," *J.Applied Microbiology,* **45**, 347-356.

Parkin, G.F. and R.E. Speece, (1982) "Modeling Toxicity in Methane Fermentation Systems," *J. Environmental Engineering Division,* **108**, 515-531.

Parkin, G.F., N.A. Lynch, WC. Kuo, E.L. Van Keuren, and S.K., Bhattacharya (1990) "Interactions between Sulfate Reducers and Methanogens Fed Acetate and Propionate," *J. Water Pollutions Control Federation,* **62**, 6, 780-788

Perkins, R.J. (1989) *Onsite Wastewater Disposal*, National Environmental Health Association, Lewis Publishers Inc., Ann Arbor, MI.

Philip, H., S. Maunoir, A. Rambaud, and S. Philippi (1993) "Septic Tank Sludge: Accumulation Rate Biochemical Characteristics," Proceedings of the Second International Specialized Conference on Design and Operation of Small Wastewater Treatment Plants, Norway.

Rittmann, B.E. and P.L. McCarty (2001) *Environmental Biotechnology: Principles and Applications,* McGraw-Hill, Inc. New York.

Roy, C. and J.-P. Dubé (1994) "*A Recirculating Gravel Filter for Cold Climates,*" Proceedings of the Seventh International Symposium on Individual and Small Community Systems, American Society of Agricultural Engineers, St. Joseph, MI.

Sahely, H.R., H.L. Maclean, H.D Monteith, and D.M. Bagley (2006) "Comparison of On-site and Upstream Greenhouse Gas Emissions from Canadian municipal wastewater treatment facilities," *J. Environ. Engr. Sci.,* **5**, 405-415.

San Luis Obispo County (2008) *Technical Memorandum, Septage Receiving Station Options*, Los Osos Wastewater Project Development. Los Osos, CA.

Sasse, L. (1998) DEWATS, Decentralized Wastewater Treatment in Developing Countries, Borda, Delhi, India.

Schönheit, P., J.K. Kristjansson, and R.K. Thauer (1982) "Kinetic Mechanism for the Ability of Sulfate Reducers to Out-Compete Methanogens for Acetate," *Arch. Microbiol.,* **132**, 285-288.

Seabloom, R.W., T.R. Bounds, and T.L. Loudon. (2005). Septic Tanks Text. in (M.A. Gross and N.E. Deal, eds.) "University Curriculum Development for Decentralized Wastewater Management," National Decentralized Water Resources Capacity Development Project. University of Arkansas, Fayetteville, AR.

SLOC (2008) San Luis Obispo County septic tank regulation.

Speece, R.E. (1996) *Anaerobic Biotechnology for Industrial Wastewaters.* Archae Press, Nashville, TN.

Tabachow, R.M., P.A. Roelle, J. Peirce, and V.P. Aneja (2002). "Soil Nitric Oxide Emissions: Lab and Field Measurements Comparison," *Environ. Engr. Sci.,* **19**, 4.

Tchobanoglous, G. and E.D. Schroeder (1985) *Water Quality: Characteristics, Modeling and Modification.* Addison-Wesley Company, Inc., Menlo Park, CA.

Tchobanolgous, G. and F.L. Burton (1991) *Wastewater Engineering: Treatment, Disposal, and Reuse*, 3rd ed., McGraw-Hill, Inc., New York.

Tchobanoglous, G., F.L. Burton, and H.D. Stensel (2003) *Wastewater Engineering: Treatment and Reuse*, 4th ed., Metcalf and Eddy, Inc., McGraw-Hill Book Company, New York.

U.S. EPA (1993), *The Determination of Chemical Oxygen Demand by Semi-Automated Colorimetry,* EPA Method 410.4, Cincinnati, OH.

U.S. EPA (1994) *Guide to Septage Treatment and Disposal,* EPA/625/R-94/002. Washingtion, D.C.

U.S. EPA (1999) *Quantification of Methane Emissions and Discussion of Nitrous Oxide and Ammonia Emissions from Septic Tanks, Latrines and Stagnant Sewers in the World,* EPA-600/R-99-089. Springfield, VA.

U.S. EPA (2000) *Decentralized Systems Technology Fact Sheet. Septic System Tank*, EPA-832-F-00-040. Washington, D.C. September, 2000.

U.S. EPA (2002) *Onsite Wastewater Treatment Systems Manual*, EPA-625/R-00/008, February, 2002.

U.S. EPA (2005) *A Homeowner's Guide to Septic Systems*, EPA-832-B-02-005. U.S. Environmental Protection Agency, Washington, D.C.

U.S. EPA (2009) *Inventory of U.S. Greenhouse Gas Emissions and Sinks: 1990-2007*, EPA 430-R-09-004, U.S. Environmental Protection Agency, Washington, D.C.

U.S. EPA (2010) *Inventory of U.S. Greenhouse Gas Emissions and Sinks: 1990-2008*, EPA 430-R-10-006, U.S. Environmental Protection Agency, Washington, D.C.

U.S. EPA Region 9 (2003) California Wastewater Training and Research Center, 2003. Status Report, Onsite Wastewater Treatment Systems in California. California State University, Chico, CA.

U.S. PHS (1957) *Manual of Septic Tank Practice*, U.S. Public Health Service, Washington, DC.

Walker, H.B. and R.H. Driftmier (1929) "Studies of the Septic Tank Method of Sewage Disposal for Isolated Homes," *J. Ag. Engr.*, **10**, 8, 9, and 10.

Weibel, S.R., C.P. Straub, and S.A. Thoman (1949) *Studies on Household Sewage Disposal Systems*, Part I. US Public Health Service, Cincinnati, OH.

Weibel, S.R., C.P. Straub, and S.A. Thoman (1955) *Studies on Household Sewage Disposal Systems*, Part III. US Public Health Service, Cincinnati, OH.

Wilhelm, S.R., S.L. Schiff, and J.A. Cherry (1994) "*Biogeochemical Evolution of Domestic Wastewater in Septic Systems: 1. Conceptual Model," J.Ground Water,* ***32****, 6, 905-916.*

Winfrey, M.R. and J.G. Zeikus (1977) "Effect of Sulfate on Carbon and Electron Flow During Microbial Methanogenesis in Freshwater Sediments," *J. Applied and Environmental Microbiology, **33**, 2, 275-281.*

Winneberger, J.H.T. (1984) *Septic-tank Systems A Consultant's Toolkit. Volume II- The Septic Tank*. The Butterworth Group, Ann Arbor, MI.

http://cfpub.epa.gov/owm/septic/search.cfm#glossary. Accessed on 08/04/09.

http://www.epa.gov/ncea/iris/subst/0061.htm, Last updated on Thursday, July 9th, 2009. Accessed on 08/19/09.

http://www.epa.gov/nitrousoxide/sources.html, Last updated on Thursday, July 20th, 2009. Accessed on 11/02/09.

IWA Publishing's authorised EU representative for General Product Safety Regulations is Diane D'Arras, 15 rue Duret, 75116 Paris, France, e-mail: safety@iwap.co.uk.

Printed and bound by CPI Group (UK) Ltd, Croydon, CR0 4YY
12/05/2026
02108865-0013